Building Mathematics Learning Communities

Improving Outcomes in Urban High Schools

Building Mathematics Learning Communities

Improving Outcomes in Urban High Schools

Erica N. Walker

Foreword by Bob Moses

TEACHERS
COLLEGE
PRESS

Teachers College, Columbia University
New York and London

Published by Teachers College Press, 1234 Amsterdam Avenue, New York, NY 10027

Elements of Chapter 5 first appeared in "The Structure and Culture of Developing a Mathematics Tutoring Collaborative in an Urban High School," by E. N. Walker, 2007, in *The High School Journal, 91*(1), 57–67. Used with permission.

Elements of Chapter 5 first appeared in "Preservice Teachers' Perceptions of Mathematics Education in Urban Schools," by E. N. Walker, 2007, in *The Urban Review, 39*(5), 519–540. Used with permission from Springer Science+Business Media.

Library of Congress Cataloging-in-Publication Data

Walker, Erica N.
  Building mathematics learning communities : improving outcomes in urban high
    schools / Erica N. Walker ; foreword by Bob Moses.
      p. cm.
  Includes bibliographical references and index.
  ISBN 978-0-8077-5328-6 (pbk. : alk. paper)
    1. Mathematics—Study and teaching (Secondary) 2. Education, Urban. I. Title.
  QA11.2.W345 2012
  510.71'21732—dc23   2012009269

ISBN 978-0-8077-5328-6 (paper)

Printed on acid-free paper
Manufactured in the United States of America

19 18 17 16 15 14 13 12   8 7 6 5 4 3 2 1

For my niece and nephews—
Justine and Juwan Walker
and
Ian Smith Jr.

Contents

Foreword *by Bob Moses*   xi

Preface   xiii

Acknowledgments   xv

Introduction   1

**1. Urban High School Students and Mathematics:
   Myths and Realities**   7

Mathematics in Popular Culture and the Media   8

Framing the Mathematics Achievement Gap   12

Teachers' Beliefs About Mathematics Teaching
and Urban Students   17

At Lowell High: Teachers' Beliefs About Their Students,
Teaching, and Mathematics   21

Students' Other Math Outcomes: Attitudes and Participation   24

At Lowell High: Students' Mathematics Attitudes,
Performance, and Behaviors   27

**2. Understanding Students' Communities and How They Support
   Mathematics Engagement and Learning**   29

Black and Latino/a Community Support for
Education and Mathematics Learning   29

At Lowell High: Family Support for Mathematics Learning   31

Peer Influences and Academic Achievement   33

At Lowell High: Students' Peer Groups and
Their Support for Mathematics   39

Conclusions   48

**3. Facilitating and Thwarting Mathematics Success
   for Urban Students**   51

National, State, and School District Policies and Practices   53

School Resources and Organizational Practices — 56

Department Policies, Student Learning, and Achievement — 61

Classroom Dynamics:
Teachers, Students, and Their Interactions — 62

Effective Mathematics Teaching — 65

At Lowell High: Students' Perceptions of Mathematics Teachers
and Their Instructional Practices — 67

**4. Engaging Urban Students' Mathematical Interests
to Promote Learning and Achievement — 71**

What Actually Happens In School Mathematics Classrooms — 74

Nonengaging Mathematics Experiences:
Lowell High Students' Voices — 77

Engaging Mathematics Experiences:
Voices of Lowell High Students and Mathematicians — 78

Bridging Out-of-School and In-School
Mathematics Learning Experiences — 82

A Framework for Student Engagement in Mathematics — 86

5. Developing a Peer Tutoring Collaborative — 88

Beginnings: Starting the Tutoring Collaborative — 89

Components of the Collaborative — 91

General Impressions and Mathematics Outcomes — 94

Characteristics of Tutor–Tutee Interactions — 96

Challenges to Traditional Hierarchies:
Collaborations Among Participants — 99

Creating a Collaborative Space for Doing Mathematics — 106

Preservice Teachers' Knowledge and Beliefs
about Urban High Schools and Students — 108

Coda — 110

6. Conclusions — 112

A Question of Opportunity — 112

Implications for Teachers — 112

Implications for Teacher Educators — 114

Implications for Administrators and Policymakers — 115

Implications for Reconceptualizing
Student Achievement in Mathematics — 117

Appendix A: Methodological Notes   **121**

Appendix B: Sample Student Map of Influences   **124**

Appendix C: Student-Designed Recruitment Materials   **125**

Appendix D: Peer Tutor Training Materials   **126**

Notes   **129**

References   **130**

Index   **141**

About the Author   **152**

Foreword

"We The People" is not just an abstract constitutional concept. The nation's constitutional democracy requires a "We The People" educational reach. Its 20th-century educational reach sought increased public access to education and increased rates of reading and writing literacy, while its 21st-century educational reach seeks to also nurture and establish quantitative literacy.[1] But "We The People" has always required moral reach, and in this century it requires us to reach deep into the moral dilemma of our "caste system," which "finds its clearest manifestation in [our] educational system."[2]

Worthy of our attention, Erica Walker's *Building Mathematics Learning Communities* studies students willing to reach for the level of mathematical understanding required for effective membership in a 21st-century, "We The People" nation—students working the demand side of our moral dilemma. In other words, their explicit self-demand to create a student-led, active-learning peer culture constructs an implicit "We The People" demand for universal, quality public school education.

This book studies what math students do, how they do it, and what support they need to disengage themselves from our national caste system as it manifests itself through educational inequality. Worthy of our attention, *Building Mathematics Learning Communities* also introduces students worthy of the heady challenge laid down by civil rights pioneer Ella Baker:

> "In order for us as poor and oppressed people to become a part of a society that is meaningful, the system under which we now exist has to be radically changed. This means we are going to have to learn to think in radical terms. I use the term *radical* in its original meaning—getting down to and understanding the root cause. It means facing a system that does not lend itself to your needs and devising means by which you change that system. That is easier said than done."[3]

In his famous "House Divided" speech, Abraham Lincoln mused, "If we could first know where we are, and whither we are tending, we could then better judge what to do, and how to do it."[4] Worthy of our attention, *Building Mathematics Learning Communities* delivers into the enduring footprint of Lincoln's speech a way forward: Sailing the troubled global waters of the industrial age as it transitions to the information age is "where we are," and extending the educational reach of "We The People" is "whither we ought to be tending." Within this historical context, *Building Mathematics Learning Communities* not only studies the "what" and the "how," it also underscores the "who": the students who must rise to face a system that does not lend itself to their needs and devise means by which they can change that system (easier said than done, indeed).

One hundred and fifty years ago freed slaves confronted the nation with their "who-ness" and the nation's conversation about the 13th, 14th and 15th constitutional amendments—the "what" and the "how" of national citizenship—turned a corner. Let us not worry that "the house is still divided"—that "where we are and whither we are tending" is still with us; slavery, after all, enslaves us all and getting out from under "600,000 down" is always easier said than done.

Worthy of our attention, *Building Mathematics Learning Communities* warns the "we" of Lincoln's speech to pay close attention to the "who" they have thought not worthy of their attention. This is no less challenging for education today than it was for slavery in Lincoln's time. It requires a "We The People" reach.

—*Bob Moses*

NOTES

1. Orrill, R. (2001). Mathematics, numeracy, and democracy. In L. A. Steen (Ed.), *Mathematics and democracy: The case for quantitative literacy* (p. xiii-xx). Princeton, NJ: The Woodrow Wilson Fellowship Foundation.

2. Conant, J. B. (1961). *Slums and suburbs: A commentary on schools in metropolitan areas.* New York: McGraw Hill, pp. 11.

3. Quoted in Moses, R. P., & Cobb, C. E. (2001). *Radical equations: Civil rights from Mississippi to the Algebra Project.* Boston: Beacon Press, p. 3.

4. Lincoln, A. (1858, June 16). "House divided." Speech delivered at the Illinois Republican Convention, Springfield, IL. Retrieved from http://www.pbs.org/wgbh/aia/part4/4h2934t.html

Preface

I have been a lover of mathematics since I was a child, when my favorite neighbor, Kelly, would give me math problems from shiny new workbooks and tell me that I was great at math. I never heard from my parents that it was odd that I, a girl, was good at mathematics. When I went to school, I learned math mostly from teachers who taught the subject traditionally but with great enthusiasm. Looking back, I can appreciate the genius of my 6th-grade math teacher, who gave us pop quizzes that didn't look like any we had ever taken before (I still remember the multiplication maze that we had to figure out). My 9th-grade geometry teacher told us that doing proofs was extremely difficult but worth doing (and doing well), and my 11th-grade math teacher made sure my senior-year schedule included advanced placement (AP) calculus. "Why wouldn't you take AP Calculus?" she asked me when I told her that I needed only 3 years of high school math to graduate. In 12th grade, my math teacher expressed some consternation that my parents were allowing me to go to Canada as an international student ambassador for the school district and was concerned that I would miss several weeks of class. My Canadian host brother helped me learn the calculus I was missing while I was away, and one of my classmates explained what I had missed when I returned.

Because I did well in school mathematics, friends and classmates sought me out for math help, as did family members and neighbors. Thus it came as no surprise to many that I became a high school math teacher. As a teacher, I remembered all the math conversations and tutoring sessions my high school friends and classmates and I had had, and I used those experiences as a foundation for a morning peer tutoring program for, and by, students who wanted to work on math.

While I received outstanding academic preparation as a public school elementary and secondary student, an undergraduate mathematics major, a master's student in mathematics education, and finally as a doctoral candidate exploring issues of policy and mathematics education, what I remember and value just as much are my math experiences with my family, with teachers and community members, and with other young people. These experiences, along with my formal preparation, have been equally important to any claim I have to being a "math person," and have influenced how I teach, think about, and do mathematics and mathematics education research.

Consequently when I walk through city neighborhoods or through the halls of urban schools I do not see them as empty of promise, although there are plenty of stories about what these areas, their schools, and their students lack. I look for the talent that I know is there. My own experiences of growing up in a big city and attending urban public schools revealed to me that there were young people, parents, teachers, school administrators, and neighbors who were interested in mathematics and interested (and invested) in my success. When I look at our cities I expect to see the same communities committed to young people's school and mathematics success. This book is about my efforts to uncover these communities, which too often are hidden or ignored, and to encourage the teachers of young people—whether in or out of school—to not only look for, look at, and learn from communities that support mathematics success, but to help create and sustain them as well.

Acknowledgments

I wish to thank the amazing students, teachers, and administrators at Lowell High for welcoming me to their school. I also thank my students and colleagues in the Program in Mathematics at Teachers College, Columbia University, for their support of this work. I particularly wish to acknowledge the support of the Department of Mathematics, Science, and Technology, under the leadership of O. Roger Anderson, and the Office of the Provost, under the leadership of Thomas James. Support from an American Educational Research Association/Institute of Education Sciences grant was instrumental in early stages of the work.

I have been fortunate to have excellent research assistants who worked with me at all stages of my work at Lowell High; among them are Nathan Alexander (who also provided greatly appreciated research support for this book), Viveka Borum, Jill Baron, Jessica Pierre Louis, Regine Phillipeaux-Pierre, Katherine Kovarik, Mia Tulao, Maurese Richardson, Bridget Ryan, Joo Young Park, and Ana Laura Martinez. In addition, students in my courses were enthusiastic about working with and learning about teachers and students at Lowell High. I thank them for their interest and commitment. Conversations with colleagues and friends, including Bruce Vogeli, Alexander Karp, Lalitha Vasudevan, Sandra Okita, Lesley Bartlett, Robyn Brady Ince, Saeyun Lee, Hal Smith, Tanya Odom, Denise Ross, Adeyemi Stembridge, Maisha Fisher Winn, Rochelle Gutiérrez, Jacqueline Leonard, Danny Martin, Dorothy White, and Karen King, helped me to sharpen my thinking and press forward on this book. I thank Edmund Gordon and Eleanor Armour Thomas for their research support and guidance during the early stages of my career at Teachers College. I would like to especially thank Bob Moses for his vision and inspiration, and for his support for my work in mathematics education.

Sincere gratitude and appreciation go to my family (especially my mother, sister, and brother) and friends who expressed interest in my work with high school students, read article and chapter drafts, and championed this book. Immense thanks to my mathematics teachers in and out of school, including Mr. Alton Kelly and Mrs. Yvonne Pringle, for their encouragement and inspiring and skillful pedagogy.

Introduction

Building Mathematics Learning Communities is based on my research focusing on improving mathematics outcomes for urban students over the past decade, building on findings from a 4-year research project undertaken at a predominantly Black and Latino/a New York City public high school, Lowell High[1]. This book seeks to promote a richer understanding of the performance, potential, and promise of urban youth in mathematics. Much of the research and discussion about urban Black and Latino/a students and mathematics focuses on underachievement and in fact appears to presume that mathematical excellence is rare among urban students of color. Further, very little research in mathematics education draws on students' voices and experiences to develop models for mathematics improvement. This book uses a different lens: It draws on the mathematics perceptions, behaviors, and experiences of students at an urban school—both high and low achievers—to drive a discussion of how educators can build stronger mathematics communities in schools. In this book, I describe students' academic communities (Walker, 2006) and peer networks, examine how these support mathematics engagement, and provide a model for how schools and teachers might build on these networks to create collaborative and nonhierarchical mathematical communities that support student engagement and achievement in mathematics.

This book provides insight for mathematics teachers (both preservice and in-service) and urban school leaders as to how to facilitate positive interactions, engagement, and achievement in mathematics for Black and Latino/a students, who compose nearly 70% of the student population in large urban areas (National Center for Education Statistics [NCES], 2010), which serve over 7 million students. These are compelling issues, given the lower, on average, mathematics achievement of Black and Latino/a students; that most national studies find that for Black students, in particular, positive attitudes toward mathematics are not correlated with high mathematics outcomes; and the findings of many researchers that urban teachers have low expectations for their students.

High school is a particularly critical time to address these issues. High school is the time when we see students of all ethnicities increasingly become disinterested in mathematics—fewer and fewer students persist in taking classes in the advanced mathematics pipeline. In addition, the high school mathematics curriculum, as discussed in this book, is hierarchical and segmented in ways that not only obstruct

1

students' paths through high school and on to graduation, but also affect their lives for decades to come. Whether advanced mathematics courses are even offered is an issue in many schools serving urban students. And it is during high school when the mathematics courses that students take signal to colleges whether or not they are worthy of admission; and it is often during high school that critical choices about whether and when to take critical gatekeeping courses such as Algebra I are made. Navigating these choices and experiences can be difficult for students.

In my research and work as a mathematics educator over the past decade, I have sought to highlight students' voices, in particular, the voices of students of color in urban schools whose mathematics experiences, behaviors, and interests have largely gone unexplored. Much of the discourse about urban students focuses on their lack of achievement and supposed disinterest in education, school, or mathematics, but my work to date reveals a complex and compelling picture of what students think about their school and classroom experiences and, specifically, their mathematics experiences in and out of school.

Throughout this book, I argue that mathematics teaching and learning reform discussions are too often limited to a "top-down," deficit-orientation perspective—particularly in urban settings—that does not adequately consider the strengths and interests of students. Mathematics teaching and learning can and should build on students' existing networks and communities and incorporate successful mathematical practices in which students and teachers may already be engaged. In my work, I have found that the networks and communities that support student achievement in mathematics are often hidden and very rarely capitalized upon by teachers and administrators. For students to be successful in school mathematics and to develop strong mathematics communities within schools, I argue that attention must be paid to the networks and communities that support their engagement and learning in mathematics *inside* as well as *outside* of school.

I argue that three key principles must be understood for teachers and students to build strong mathematics communities in urban schools, and thus improve mathematics outcomes for students in those settings. These principles, which emerged from my work at Lowell, are supported by the extant research literature. First, I suggest that despite common portrayals, urban high school students want to be engaged in mathematics and have developed communities that support that engagement. I base this assertion and discussion on surveys and interviews with Lowell students, as well as extant research and national data describing students' positive attitudes toward mathematics and their supportive peer groups for mathematics. Second, I suggest that *teachers and administrators can inadvertently create obstacles that thwart the mathematics potential of these students*. School adults should engage in practices that best facilitate students' mathematics success, not impede it. Evidence for this discussion comes from interviews and questionnaires of Lowell teachers and preservice teachers, as well as extant research about the perceptions and practices of administrators and mathematics teachers in urban schools. Finally,

I suggest that students' existing networks need not be hidden or unused—*that teachers and administrators can learn from how students develop mathematics communities and use this knowledge to build collaborative communities that support mathematics achievement and engagement.* One example of how this might occur is the peer tutoring collaborative project I developed at Lowell High School. I use this example to describe how schools might use students' positive peer networks to drive mathematics engagement and achievement, but also posit other ideas for capitalizing on students' networks in this section of the book.

These three principles are the organizing framework for Chapters 1 through 5 in *Building Mathematics Learning Communities*. This book, unlike much of the research on urban students and mathematics, is built on a foundation of what we can learn from students' experiences with mathematics in and out of school. It describes in a practical way how we might use this knowledge to improve mathematics teaching and learning in urban schools, create and sustain communities that support and actively engage student learning, and develop cadres of students with strong and positive mathematics identities who are excited about mathematics and see themselves and are seen as talented, knowledgeable doers and users of mathematics.

Principle 1. Urban High School Students Are Interested in Mathematics

In Chapter 1, Urban High School Students and Mathematics: Myths and Realities, I describe pervasive myths about urban high school students and mathematics (namely, that these students are disinterested in mathematics and are, on the whole, lower achieving) and provide evidence that refutes these myths. I discuss how urban students view their own mathematics experiences, both in and out of school. Augmenting what we know from the literature, I use Lowell students' survey data to deeply examine students' beliefs about and experiences with mathematics and mathematics class. In addition, I discuss the relationship between these beliefs and the background characteristics of these students (such as ethnicity, gender, and prior achievement in mathematics). For example, much research in mathematics education examines the role of gender in mathematics success and failure, but very few studies—despite entreaties by mathematics education researchers—have examined gender issues within particular ethnic groups and in urban settings as these relate to mathematics. There is some evidence that the gender gap in mathematics performance, which generally favors boys nationally on standardized assessments like the SAT, is reversed for students from particular ethnic backgrounds and in certain settings. In some cases, this proved true at Lowell. This has implications for how teachers and students interact with each other, and the gendered expectations that teachers, community members, and students may have for students' mathematics success. Also included in Chapter 1 is a discussion of mathematics teachers and their beliefs about mathematics, its teaching, and their students. In Chapter 3, I revisit some of these issues to discuss how they affect instructional practice.

In Chapter 2, Understanding Students' Communities and How They Support Mathematics Engagement and Learning, I describe the communities that support students' mathematics success, beginning with their out-of-school networks (which comprise family members, significant adults, and other young people) and ending with their school-based peer communities. Within this discussion, based on extant research as well as survey and interview data from Lowell High School students, I explore links between these communities and students' ethnicity and gender. This chapter's primary goal is to challenge a prevailing myth about urban students and their peer networks: that peers contribute solely negative influences and do not necessarily facilitate academic success. Very few studies explore this question as it relates to mathematics achievement.

Building on the recent research literature that challenges notions of "oppositional schooling" for urban student and students of color, I focus on a group of Lowell's high-achieving students and describe the networks and behaviors that these students believe facilitate their mathematics success, again addressing issues of race and gender. It should be noted that these networks are wide ranging and inclusive; for example, in addition to other high-achieving students, these networks include lower-achieving students, parents who may or may not have graduated from high school, and extended family members who talk to students about the importance of mathematics and mathematics learning. I suggest that teachers and researchers must begin to think more broadly and deeply about the external (out-of-school) as well as internal mechanisms of support for students that can be used to drive improved mathematics achievement in school.

Principle 2. Schools and Teachers Can Better Facilitate Students' Mathematical Learning

In Chapter 3, Facilitating and Thwarting Mathematics Success for Urban Students, I discuss the importance of broader national, state, local, and school policies in shaping opportunities for students. It is important that teachers and administrators recognize that mathematics as a subject is unique, both in how it is viewed in U.S. society and in how it is structured, organized, and delivered in school. I argue that these twin issues—how mathematics is perceived by the public and how it is structured in schools—are related. In particular, teachers and administrators may wittingly and unwittingly perpetuate practices that limit students' access to mathematics and underscore mathematics as something only the talented few can do. Chapter 3 discusses school organizational practices and teacher expectations (e.g., R. F. Ferguson, 1998) and links these to students' mathematics classroom experiences and outcomes. Building on research that explores practices in schools that both facilitate and limit access to mathematics, I invite the reader to critique or defend teachers' and administrators' beliefs, practices, and policies about students, mathematics, and mathematics teaching and examine how these operate to either facilitate or thwart student success.

In order for mathematics teachers and others to build on students' positive peer networks and extensive academic communities to foster improved mathematics outcomes, they have to first believe that these are worthwhile and have mathematical value. They have to be willing to expand notions of readiness and potential to improve access to gatekeeping courses, such as algebra. I use data from observations of mathematics classrooms and student survey and interview data at Lowell High to ground a discussion about the ramifications of the ways in which mathematics is currently organized in most school settings for student learning.

Chapter 4, Engaging Urban Students' Mathematical Interests to Promote Learning and Achievement, discusses the research evidence that shows that for most students, the high school classroom is not engaging. In particular, in most urban schools and at Lowell High, school mathematics follows the traditional didactic instructional model, and students' classroom behaviors echo both comfort and discomfort with this model. High school mathematics reform efforts, I argue, are too often narrowly centered on curriculum and assessment, and do not address the importance and quality of the dynamic instructional opportunities that teachers could provide for learners on a day-to-day basis. In this chapter, I argue for rigorous, content-based instructional strategies that incorporate students' interests and experiences. In addition to providing research-based examples from the extant literature, I discuss the ways in which Lowell High students describe their mathematics learning in and out of school and provide examples as to how these experiences might drive instruction (as well as curriculum and assessment) that captures and enhances the richness of what students know about mathematics. I also draw upon my study of successful adults in mathematics (all PhD mathematicians) to describe mathematically engaging experiences within schools, outside of schools, and in "hybrid," or in-between, spaces. The purpose of this discussion is to elucidate a framework for how teachers might better engage students' interests in the classroom that is not dependent on a particular curriculum or type of assessment, or even a particular instructional model, but that provides access to rigorous mathematical content. Further, these kinds of enriching experiences do not have to be limited to the mathematics classroom—we can take a more expansive view of mathematics education and extend mathematics learning beyond the traditional confines of the school day and classroom. Because research suggests that students learn more mathematics when they have the opportunity to engage in rigorous mathematical talk with peers and teachers, I argue that developing ways to intensify student-student interaction—and provide opportunities for teachers to observe and learn from these interactions—are key to creating strong mathematics programs in high schools. In the next section of the book, I discuss one way that this type of rigorous teacher-student collaboration can occur: a peer tutoring collaborative project that comprised high school students, graduate students, preservice teachers, and in-service teachers.

Principle 3. Teachers and Students Can Collaboratively Develop Mathematics Communities in Schools

Chapter 5, Developing a Peer Tutoring Collaborative, describes a model for a school-based, student-led initiative that uses peer tutoring to address students' lack of engagement and achievement in mathematics. This initiative included high-achieving students as tutors, their lower-achieving peers as tutees, graduate students in mathematics education, preservice teachers, and in-service teachers. The goal of the peer tutoring collaborative was to expand students' networks beyond those of high achievers and create a supportive culture for mathematics for all students. The model was developed in part from what we learned about students' mathematics experiences at Lowell High, particularly from the academic communities of high-achieving students. The model is three pronged: (a) it suggests a site-based approach to building on existing student excellence in mathematics to drive improved student mathematics achievement; (b) it seeks to address the lack of teacher knowledge about urban students and their mathematics understanding; and (c) it aims to deepen existing mathematics knowledge, confidence, and interest among high school students. In Chapter 5, I describe the conceptual underpinnings of this model and how it is supported by research in psychology, sociology, and mathematics education. I also present sample materials developed for the program for teachers and administrators to use in their own schools.

The chapter also includes analyses of the interactions among tutors, tutees, mentors, and teachers who participated in the peer tutoring collaborative project; the mathematical discourse within those interactions; and the hierarchical and collaborative relationships between teachers (in-service and preservice) and students that emerged over time. I also focus on preservice teachers' developing perceptions of the mathematics abilities of urban students. In addition, I present evidence drawn from interviews of teachers and students that describes their learning of mathematics content (for high school students) and pedagogical strategies (for teachers and high school students). Evidence that demonstrates students' mathematics improvement is drawn from participating students' test scores and mathematics course grades.

In the final chapter, I call for schools, teachers, parents, and students to actively strive to build mathematics communities in urban schools. I argue for a broader conception of mathematics teaching and learning in urban schools, and suggest that administrators, teachers, and students develop collaborative communities that build on students' strengths to drive improved achievement, rather than solely focusing on students' "deficits," drawing from programs that work. I suggest that the development of a peer tutoring collaborative (of students, teachers, mentors, and preservice teachers) is one approach that can spur new thinking about students' mathematical potential as well as mathematics pedagogy. In addition, I posit additional strategies based on students' peer communities that administrators, teachers, and other concerned adults might use to foster students' mathematics achievement.

CHAPTER 1

Urban High School Students and Mathematics: Myths and Realities

Youth, and specifically high school students, are often constructed in media in multiple ways that portray them as disaffected and uninterested in school. However, urban adolescents—particularly Black and Latino/a students—are unique in that they are most often and popularly portrayed in ways that diminish attention to them as intellectuals (A. A. Ferguson, 2001; Pimentel, 2010). They are portrayed as commonly engaged in criminal behavior, resistant to education and learning, and disrespectful to adults. As Morrell (2008) describes, "One only need[s] to browse the newspapers of any major city, or watch popular television shows or major new film releases to gain a sense of the popular messages circulating about the urban poor and people of color" (p. 6). These depictions largely construct urban students, particularly Black and Latino/a teenagers, in ways that diminish the importance of attending to their academic and intellectual selves. Because popular culture "shapes and reflects the beliefs of Americans," these depictions of urban schools, classrooms, and students can "reflect and shape the assumptions with which preservice teachers enter urban classrooms" (Grant, 2002, p. 78). In my work, I have discovered that such depictions—whether in the press or in popular culture—filter and factor strongly into the perceptions that preservice teachers and in-service teachers hold of their own students, and how their families and friends construct urban youth for them. Said one teacher, Abby,

> I see—a lot of people have seen—those urban high school movies that Hollywood puts out that kids are running around and drugs are all over, and guns, and that I didn't see. Who knows if it's there? Maybe, maybe not, who knows. But it was one of those things where, I was, I don't want to say pleasantly surprised. Maybe I was just snapped back to reality. . . .
>   When I came down here, I was like yeah, I'm going to take a job in a Harlem school. Really, friends of mine back in Connecticut, and it sounds hoity-toity, but they'd ask, "Have you lost your mind, what are you doing going to Harlem to work? Are you nuts?" . . . I won't lie, it was a little

nerve-wracking but it's one of those things where it's a neighborhood; you know, they have kids; and kids need to be taught; they have teachers. . . . Sometimes they get forgotten about, who knows. But I [first] felt like, when we go there the kids are going to be horrible and the teachers are going to have no clue what they're doing and it's just gonna be, just mayhem. It's nothing of the sort.

Understanding urban high school students and their mathematics performance in school requires that we contest common and popular depictions of urban students and recognize that those depictions affect how they are viewed and treated in school by school adults. In addition, mathematics is a special discipline, constructed in a particular way by media and society like no other. The traditional depictions of mathematics—as the purview of nerdy, socially outcast men, largely—have an impact on how young people and adults view the subject and the people who do mathematics well. For some students, being a "mathematics person" means subverting that part of one's identity that makes a person interesting and cool (Boaler & Greeno, 2000). This chapter examines the myths and realities about urban high school students and mathematics, and how those myths and realities affect issues of mathematics teaching and learning.

MATHEMATICS IN POPULAR CULTURE AND THE MEDIA

Unfortunately the dominant depiction of mathematics in media and popular culture in the United States is that of a discipline that only a select few people do. There is a prevalent view that people who do well in mathematics do so "naturally." Consequently, unlike other disciplines that we believe require hard work—good writing can be developed, for example—our societal emphasis on mathematics as a difficult subject in which we expect few people to do well hampers our development of mathematically proficient people (Moses & Cobb, 2001) of all backgrounds. We accept underachievement in mathematics as a natural state of affairs, unlike the prevailing expectation in some other countries that all students "master a level of mathematical understanding equivalent to that attained by only our best students" (Vetter, 1994, p. 7).

This perception of mathematics as the purview of a select few is particularly damning for students who are considered to be outside of the mainstream. And indeed, perceptions of people are also defined by where and how they are absent. When we do see representations of people doing mathematics, Black and Latino/a people are conspicuously absent. In fact, the people who are doing mathematics are usually construed as a little odd, or weird, or with substantial mental problems. While some might argue that it would be no great thing to have Blacks and Latino/as represented in such a negative light, their absence from representations of these careers in media—and their usual presence in limited (sports, entertainment) and problematic spheres—filters into the societal consciousness. (As one of my graduate students once said, "I would kill for a positive stereotype about Black kids that didn't involve sports or entertainment.") A recent advertisement for Intel, seen in *Sports Illustrated*, makes this very clear. With a graphic of photographs of high school–aged students labeled as "overachievers," as though in a high school yearbook, overlaid with mathematical equations, not one of the twelve or so students represented appeared to be Black or Latino/a. Such an advertisement sends a clear message: "Overachievers" are predominantly White, male, or of Asian descent. When that image is juxtaposed with the number of images of Black athletes in the magazine, the message sent is startling in its positioning of Blacks as physically gifted and Whites and Asians as intellectually so.

It is certainly true that urban youth and Blacks and Latino/as are not the only ones missing from positive discourses about mathematics. Women's positioning with regard to mathematics is also problematic. The furor that emerged from a talking Barbie doll (Teen Talk Barbie) that said, "Math class is tough!" in 1992 and former Harvard University president Lawrence Summers's comments in 2005 about women's underrepresentation in the sciences being largely due to the "differential availability of aptitude at the highest levels" between men and women were largely indicative of the public's recognition that words and images have power and influence girls' and women's (and boys' and men's) constructions of themselves as potential mathematics doers. Even a film like *Salt* (2010), an action movie and spy thriller in which the female protagonist is a heralded operative who works for the CIA, contributes to societal perceptions of mathematics as the purview of the male. In a scene in which the protagonist, Salt, has demonstrated her capability to escape and survive, she pauses for a conversation with a young neighbor. Salt asks what she's doing, and the little girl says she is doing her math homework. "I hate math," says the super-capable spy.

Researchers such as Picker and Berry (2001) and Moreau, Mendick, and Epstein (2009) have written extensively about television shows (e.g., *NUMB3RS*), movies (*A Beautiful Mind, Good Will Hunting*), and books (*The Curious Incident of the Dog in the Night-time*) that continue to portray math as something that "odd" men and boys do or downplay women's talent in math. Moreau and colleagues (2009) reported that the students in their study (secondary school students as well as mathematics and humanities undergraduates) had several stereotypes of mathematicians. Students had specific ideas about how mathematicians look and how they act. In particular, mathematics "was written on the body" (p. 3): mathematicians had certain physical characteristics—they were odd-looking, with wild hair, for example—and they were largely represented as White and male. Students also felt that mathematicians had some mental and social issues, lived lives that were dominated by mathematics, and rarely did anything else. In similar studies with teachers, researchers have found that teachers describe mathematicians similarly.

A study by Cirillo and Herbel-Eisenmann (2011) found that teachers often position mathematicians as having "special talents that 'real' people do not have, and therefore the average person cannot do mathematics" (p. 74). In addition, teachers may construe some positive behaviors of mathematicians as negative (for example, seeking "efficient" ways to solve problems was characterized as "laziness"). Further, the work of mathematicians is construed as isolating and rooted in historical problems, rather than the social and collaborative activity it often is (Burton, 1999), and the important use of mathematics in solving contemporary problems is often underemphasized in mathematics classes. Too often, even when students and teachers question the images of mathematicians in media, they are "unable to refer to some alternative representations of mathematicians, possibly because of the lack of these available within their experiences of school mathematics and popular culture" (Moreau et al., 2009, p. 3–4).

There are at least two alternate media representations that present Black or Latino/a youth as capable of mathematics. *Stand and Deliver*, a movie based on the true story of Jaime Escalante, a mathematics teacher at East Los Angeles's Garfield High, and *A Different World*, a television show that portrayed the lives of predominantly Black students attending a fictional historically Black college, Hillman, simultaneously reify and shatter stereotypes about students of color and mathematics.

Jaime Escalante's success at Garfield High School in East Los Angeles—increasing the numbers of students taking AP calculus courses and the AP examinations for college credit—is well known and undeniable. At one point in the 1980s, more than 25% of the Latino/a students taking the AP calculus examination in the United States were from Escalante's mathematics program at Garfield. Several articles and research reports (e.g., Escalante & Dirmann, 1990; Lanier, 2010; Meek, 1989) have described Escalante's work with the predominantly working-class and poor Latino/a student population. Despite these reports and Escalante's own narrative (Escalante & Dirmann, 1990) of his work with and for students, *Stand and Deliver*, as a piece of media, stands as the definitive portrait of Escalante's work. The students in *Stand and Deliver* are portrayed largely as gang members and unruly and disrespectful students whose parents are less committed to their children's education (Grant, 2002; Pimentel, 2010) before Escalante works his magic on them; his mathematics teaching becomes a "vehicle for his students' transformation from hooligans and daydreamers" (Appelbaum, 1995, p. 80). Because their eventual success is regarded by the powers that be (the Educational Testing Service [ETS]) as improbable, their credibility (and Escalante's teaching) are impugned before students and teacher are vindicated by students' demonstrating repeated success on the AP examination. While the film eventually goes to lengths to portray the students as knowledgeable and engaged in mathematics, it arguably goes to even greater lengths to portray them in the stilted stereotypical images of urban Latino/as (Pimentel, 2010). The viewers of the movie are left with the message that it is a surprise that these students can excel in mathematics. The film itself might not have been made—despite the local Spanish-language press's attention to Garfield's success over the years, it wasn't until the ETS testing furor in 1982 that Escalante and Garfield gained widespread attention in the media, notably in the *Los Angeles Times* (Appelbaum, 1995). Further, as many have observed, the film omits that it actually took Escalante 8 years to build up the program to the success recounted in the movie, that for many years he had strong administrative support and, later, financial support from private foundations, and that there was a team of teachers who bought into his philosophy and that together they changed the culture of mathematics in the school (Escalante & Dirmann, 1990; Lanier, 2010). Appelbaum (1995) describes the media rhetoric and portrayals of Escalante as part of a public mythology about "superteachers" who can accomplish miracles in "tough" environments, supposedly "against the odds." Pimentel notes that her teacher education students, when reviewing the film, believed that the fact that such a movie was made at all about Latino/as taking and passing the AP exam and the presentation of this as a novelty speaks volumes about societal expectations for urban Latino/as.

Shockingly, and in contrast to *Stand and Deliver*, very little research has been done examining the possible cultural impact of *A Different World* (1987–1993) on adolescents' educational aspirations. *A Different World* was the first television show to regularly depict young Black people in college classrooms. While there are numerous essays and commentaries in the popular press that suggest that this television show was responsible for an increase in Black adolescents' interest in college and attending historically Black colleges and universities (during the 1990s in particular), there is little empirical research on this issue. Developed and produced by Bill Cosby, well known for his interest in education, *A Different World* was a spin-off of the immensely popular *The Cosby Show* (1984–1992), following one of its main characters, Denise Huxtable, off to college. (One might argue that the experiences of the eldest daughter, Sondra Huxtable, at Princeton might have been worth chronicling as well). But rather than focus on *A Different World*'s many positive images of Black students—and its portrayal of Blacks of multiple and complex backgrounds and experiences (Gray, 1995)—I wish to focus on one mind-bending, subversive character: Dwayne Wayne, an uber-nerd from Philadelphia who not only became cool and got the girl but also was shown doing math, loving math, earning his PhD in math, and embarking on what appeared to be a research career in mathematics.

What is brilliant about the characterization of Dwayne Wayne is that the *Different World* writers allowed him to shift and mature from a nerdy, socially clueless yet caring misfit to an intellectual, attractive, and urbane adult who showed America what we had not seen on the small screen: a mathematically talented Black American. (Steve Urkel, on *Family Matters*, whose interests were primarily scientific, was soon to follow. But, unlike Dwayne Wayne, he was commonly the butt of jokes and remained helplessly a nerd, despite the development of a story line that allowed his cool but callous "alter ego," Stefan Urquelle, to emerge. Also, *Family Matters* lacked any race-conscious context, while *A Different World* was firmly situated in the physical and cultural milieu of a historically Black college.) But Dwayne Wayne was unique in that his personality and humanity were not at odds with his mathematical abilities; he did not have to literally or figuratively become a different person in order to excel.

One young mathematician interviewed for my research project on Black American mathematicians describes the impact of Dwayne Wayne on her decision to become a mathematics major in college:

> Did you watch *A Different World*? Anyway, I was in love with Dwayne Wayne. . . . I thought I was so cool. Like, "Ooh. It's cool that he majored in math." I had this glorified image of going to an HBCU [historically Black college or university] and majoring in math. I thought I was cool like Dwayne Wayne. Part of me was just like, "I guess that will be okay," because I saw it. And I was just like, "Okay. Well, I saw it on TV so it's not crazy."

Sadly, characters like Dwayne Wayne—a young Black man from an urban area, smart, interested in mathematics, in college, and not a troublemaker—are few and far between in media today. Instead, we see that young people who are Black and Latino/a, particularly in cities, are constructed in ways that reify destructive stereotypes and diminish their humanity. Unfortunately, in schools, this means that many educators have images of urban students that position them as uninterested in school and with limited intellectual potential.

I share the example of Dwayne Wayne and the mathematician because it reminds us that narratives have power and, further, that positive narratives need to be shared not just as counternarratives, but also so that they might be used to interrupt dominant narratives about achievement and who can and is expected to achieve at high levels. This is particularly true of the narrative about the mathematics achievement gap in the United States, especially as it pertains to urban schools and students.

FRAMING THE MATHEMATICS ACHIEVEMENT GAP

> It is interesting to note that the portrayal of African American students as poor performing is aligned with the broader stereotypes about African Americans as unmotivated and unintelligent. (Nasir, Atukpawu, O'Connor, Davis, Wischnia, & Tsang, 2009, p. 232)

While there is no genetic evidence that students from underrepresented groups are inherently unable to do well in mathematics (Jencks & Phillips, 1998), the focus on achievement differences between groups—that Asian American and White students perform better on standardized mathematics assessments than Black and Latino/a students, on average—often engenders assumptions about individual students' abilities and potential and marks entire groups of students as underachievers without critically analyzing the reasons for high and low performance. Further, the methods of identifying mathematically talented students—largely based on standardized test scores in most schools—unfortunately help to perpetuate the myth that some groups of students are "naturally" better at mathematics than others.

Performance gaps appear early in elementary school, persist throughout students' educational careers, and indeed, for some groups, widen as students progress through school[1]. For example, the performance gap between Black students and White students is narrowest in elementary school and widens as students progress through school. In mathematics Asian American students on average have traditionally outperformed White students, who in turn outperform Latino/a, Black, and Native American students. For example, in 2009, 52% of Asian American 12th graders scored at the "proficient" level or above on the National Assessment of Educational Progress (NAEP) mathematics assessment, compared with 33% of White students, 12% of Native Americans, 11% of Latino/as, and 6% of Blacks (NCES, 2009). Although gaps in achievement among ethnic groups have narrowed during particular time periods since the initial NAEP administration in 1971—most notably in 1988 for 17-year-olds (Jencks & Phillips, 1998; Ladson-Billings, 2007)—gaps between the Black, Latino/a, and Native student population and the Asian and White student population persist. Unfortunately, the NCES does not routinely carry out and report examinations of interactions between demographic categories—for example, between gender and race, or socioeconomic status and race—although some researchers (e.g., Strutchens, Lubienski, McGraw, & Westbrook, 2004) have done this for specific NAEP years. (I will discuss the gender/race interaction later in this chapter).

But reasons for these differences in performance are complex and intricately related. Substantial economic disparities and differing levels of parental and "grandparental" education reflect effects of both lingering and ongoing racial and ethnic discrimination and are significant reasons for the continued performance gaps. Patterns of housing segregation and discrimination are linked to rigid segregation and inequitable resources in urban schools. Evidence from the NCES in 2009 reveals that NAEP scores are higher, on average, in low-poverty versus high-poverty schools, and suburban schools (average score 157) versus town schools and rural schools (151). There was no significant difference between NAEP scores in city schools (152) and those in suburban schools. But "mean mathematics scores [on the 2002 NAEP] for students attending urban high-poverty schools are considerably lower when compared with students attending schools in more prosperous districts" (McKinney, Chappell, Berry, & Hickman, 2009, p. 279). What we do know about urban school issues (Anyon, 1997, 2005) is that funding inequities and teacher shortages often result in urban school students being taught mathematics by teachers who are less qualified and more inexperienced than those who teach in suburban schools[2]. In addition, urban school students (who are overwhelmingly Black and Latino/a) often receive mathematics instruction centered on basic skills and repetition, rather than instruction that provides them with opportunities to learn and exercise higher-order thinking skills. When computers are present in their schools, for example, they may be more likely to be used for basic skills rather than for mathematics exploration or enrichment. Although learning basic skills is necessary, this should not be the upper limit of what is expected from Black, Latino/a, and Native students. Finally, beginning in elementary school, Black and Latino/a children are more likely to attend schools with fewer monetary, curricular, and staff resources than are their White and Asian counterparts. These differences result in very real obstacles for Black and Latino/a high academic performance.

Studies of elementary school achievement have shown that differences in academic outcomes are almost entirely explained by the quality of instruction that students receive, and not by race or socioeconomic status (Darling-Hammond, 1995). Further, when Blacks, Latino/as, and Native Americans attend predominantly White elementary schools, they are less likely than Asian and White students to be placed in high-ability groups (Secada, 1992) or gifted programs (D. Y. Ford & Harris, 1999)—opportunities necessary to foster strong mathematics performance.

Obviously, the mathematics learning opportunities received in elementary school are important, but the middle school years mark a critical milestone in the educational careers of students. It is here that an important hierarchy emerges, extending beyond any ability grouping in elementary school. In middle school, "the hierarchical nature of math course taking and its accompanying structure of prerequisites" (Riegle-Crumb, 2006, p. 103) means that if one does not take Algebra I, a critical gatekeeping course, early enough in secondary school, in most schools it is difficult to accelerate or "catch up" to more advanced mathematics courses. Unlike with other disciplines like English and social studies, there are far fewer options to gain access to advanced material in the mathematics curriculum. Tracking in mathematics, unlike in other disciplines, operates on at least two levels: both whether one enters the advanced mathematics pipeline at all, and when (Walker, 2001).

Student entry into the college preparatory mathematics pipeline (the sequence of courses that includes Geometry, Algebra II, Trigonometry, Precalculus and Calculus) through Algebra I, an important gatekeeper course, is based on a sometimes arbitrary system of course placement. Students whose parents are well connected, affluent, and highly educated and who know how to "work the system" are more likely to be placed in high-level mathematics in middle school, regardless of their test scores (Useem, 1992). In other words, merit is not sufficient—for example, even when Black students have higher scores than those of White students, the former may be placed in lower-level math classes (Oakes, 1995). Not only are Black and Latino/a students less likely than White and Asian students to enroll in algebra "early" (in 8th grade or before), but they are also less likely to enroll in algebra in 9th grade. Consequently, non-Asian students of color are consistently underrepresented in the courses that make up the "advanced" part of the mathematics pipeline in high school. For example, 21.8% of Native American, 41.7% of Black, and 34.3% of Latino/a high school graduates take "advanced academic" mathematics courses like trigonometry, precalculus, and calculus in high school, compared with 54.3% of Whites and 69.1% of Asians (NCES, 2007). The figures for calculus specifically are even more disheartening: only 5.6% of Native American, 4.7% of Black, and 6.8% of Latino/a high school graduates were enrolled compared to 16.0% of Whites and 33.4% of Asian American graduates. There may be student interest in these courses, but in many urban schools calculus is not even offered (Gutiérrez, 2000; Oakes, 1990; Werkema & Case, 2005). This disparity in opportunity is made starkly clear by Werkema and Case (2005):

> An honor student attending a public school in an affluent community is likely to graduate with 4 years of lab science, a mathematics sequence culminating in calculus, 3 or more years of a foreign language, and a handful of advanced placement courses equivalent to college credits. In contrast a promising student at an urban school may find only 1 year of lab science offered, a mathematics sequence that tops out at advanced algebra, and no foreign language or AP courses available. (p. 498)

Further, even when there are advanced classes in urban schools with the labels of *AP* and *honors*, "the type of preparation that they offer their students [may be] far from equal" (p. 499).

Because Black, Latino/a, and Native American students drop out of mathematics earlier and at rates higher than those of their White and Asian counterparts, the secondary mathematics classroom has been called "one of the most segregated places in American society" (Stiff & Harvey, 1988, p. 152).[3] Even after prior mathematics achievement and socioeconomic status are taken into account, Black students are less likely than their White counterparts to persist in advanced mathematics.[4] This segregation of students in lower-level mathematics classes throughout their high school careers has a significant impact on their mathematics performance—evidence from the NAEP shows that students who take more, and more advanced, mathematics courses score higher than their counterparts on the exam. Consequently, other enriching experiences—mathematics clubs, competitions, and college programs targeting strong mathematics students—are often not available to students of color, since college preparatory mathematics classes are usually the pools from which these students are drawn. Thus, at every stage where opportunities for developing mathematically talented students might occur, Black, Latino/a, and Native American students are at a disadvantage compared with their Asian and White American counterparts. In addition, these course-taking disparities have dire consequences for students' future college-going possibilities and career success. Nasir and colleagues (2009) point out that the system of mathematics education stratifies access to and positioning in the field of mathematics. Thus we "must prepare urban students to simultaneously challenge existing hierarchies of knowledge and to be competitive in a system that relies on such hierarchies" (p. 228).

Many have argued that how we frame achievement disparities matters a great deal. In particular, the dominant achievement gap discourse, focused on under-achievement and failure of some groups, does not allow for the possibility of African American or Latino/a excellence in mathematics. This is, in part, why the main-stream media discussion that eventually developed around Jaime Escalante's success at Garfield High after years of success revolved around "how a teacher had rounded up so many students with Hispanic surnames to take the AP test in the first place, whether they had cheated or not" (Appelbaum, 1995, p. 75). When Latino/a and Black students exhibit high achievement in mathematics, then, they may continue to be positioned as low-achieving despite their success. This has implications for how they are granted access, or deemed worthy of access, to mathematics resources. The fact that there are documented instances of students of color who excel being inappropriately steered into low-track classrooms is one example, but it is important to note that the ignoring of achievement by students of color is not a new story. As Wiggan (2007) describes, when a researcher conducted a study in Washington, DC, in 1897 and Black students outperformed their White counterparts on an examina-tion, the researcher "asserted that Black students were intellectually deficient" and critics of the test "called for a revision . . . because the outcome did not support the prevailing belief in White superiority" (p. 312).

Finally, when we speak of achievement disparities without discussing the historical contexts for education in the United States that contributed to centuries of disparate attention to education for large segments of society—well into the 20th century—and ignore contemporary contexts that continue to stratify edu-cational opportunity for students in urban schools, we neglect a critical part of the story, as Flores (2007), Martin (2009b), and Ladson-Billings (2007) describe. Writing about the media myth around Escalante and Garfield High, Appelbaum (1995) notes:

> The Escalante myth constructed particular ways of talking about ethnicity and class. . . . People in East Los Angeles, as in many large urban centers, were overwhelm-ingly Mexican American; most had little access to power, goods, or services, because they were poor or lower middle class. According to the story they needed to seize knowledge, and prove that they could do whatever anyone else (i.e., any middle-class Anglo-American) could do; they should have made a place for themselves within the access-rich professional and educated world (e.g., by going to college). That is, no special arrangements were needed to adjust access to knowledge and its distribution, beyond relying on a pool of superteachers who, through their own personal mission, enabled the window of access to be widened. (pp. 84–85)

Stand and Deliver and other films about urban schooling, perhaps unwittingly, present narratives that disregard the "powerful political, economic, and social forces at play in urban schools and fram[e] the teacher as a savior who . . . rescues the child from a dangerous environment" (Grant, 2002, p. 85) and may also support the notion of weakening positive ties between students and their home communities.

Martin (2009b), Hilliard (2003), and others suggest that the achievement gap discussion permits people to view students within a racialized hierarchy, and Black and Latino/a students especially, who are usually positioned at the bottom of this hierarchy, suffer for it. By focusing on Black and Latino/a students as the "problem," it detracts attention from the need to address societal and institutional structures that perpetuate inequities that are directly linked to the gaps we see. Education research that questions the focus on discussions of Black student "failure" positions the achievement gap as an opportunity gap that reflects historic and contemporary patterns of discrimination and inequity (Flores, 2007; Ladson-Billings, 2007). Gutiérrez (2008) likens uncritical discussions of the achievement gap as nothing but a "gap-gazing fetish" that does little to improve education or educational outcomes for Blacks or Latinos. And Ladson-Billings (2006) frames the achievement gap as evidence of an education debt that we have accumulated as a nation and must repay. These alternative framings of the achievement gap, unfortunately, lack critical currency within mainstream media and many educational policy circles. And most unfortunately, there is substantial evidence that this kind of contextual framing of the achievement gap does not occur in the minds of many teachers— both those who are already in the classroom and those who are studying to become classroom teachers.

TEACHERS' BELIEFS ABOUT MATHEMATICS TEACHING AND URBAN STUDENTS

If we are to better understand students' experiences with school mathematics, it is critical that we understand a central factor in their experience: the mathematics teachers with whom they interact every day. (Philipp, 2007, p. 75)

Teachers themselves are products of societal messages about mathematics, the competing schools of thought about how it should be taught, and to whom mathematics should be taught. Teachers' expectations for students' behaviors and performance, then, are filtered through their own exposure to these messages. The literature on teachers' expectations of students, particularly as it relates to demographic characteristics like race/ethnicity, gender, and socioeconomic status, reveals that teachers generally have lower expectations of Black, Latino/a, and poor students than of

White and more affluent students (R. F. Ferguson, 1998). Many researchers have posited that these beliefs—and attendant behaviors—over time contribute to a cumulative depressing of students' opportunity to learn and thus, their academic achievement.

Despite the popular view that mathematics is a "universal language," the assumption that math is culture free and color-blind can mean that "little attention is paid to students' cultural or linguistic backgrounds" in mathematics classrooms (Gutiérrez, 2002, p. 1048) despite evidence that these backgrounds can and do have an impact on mathematics classroom discourse and dynamics (Cobb & Hodge, 2002). Horn (2008), Lubienski (2000), and others have shown that teachers' conceptions of mathematics are related to how they construct and position students, often along socioeconomic status, racial/ethnic, or perceived ability (low-track versus high-track, for example) lines. Research on teachers' beliefs about ethnic minority students "has found that cultural differences between students and teachers also contribute to teachers' unfairly low opinion of these students' academic abilities" (Gandara & Contreras, 2009, p. 104). Further, students' cultural and linguistic backgrounds may affect teachers' instructional practices and expectations of students negatively. For example, for Latino students for whom English is not their first language, teachers may expect that a lack of English proficiency may reflect low mathematical skills. In addition, it means, unfortunately, that Black and Latino/a students' cultural backgrounds are not viewed as contributing strengths to their mathematics performance despite evidence to the contrary (Gutiérrez, 1999; Martin, 2009a).

More specifically, the interplay between the dominant achievement gap discourse and teachers' own attitudes and practice contributes to how students are constructed and positioned in school mathematics. While Weis and Fine (2000) and others have described extensively how urban youth are positioned in particular and often negative ways in school more generally, recent work by Hand (2010), Horn (2008), Jackson (2009), and Spencer (2009) explores the mathematics classroom as a space for the construction/deconstruction of students as engaged or disengaged participants in the mathematics classroom. In their work, they describe how mathematics teachers construct students, on occasion along racial, gender, or socioeconomic status lines. For example, in a case study of two teachers, Spencer (2009) notes that they operated within three "lines of discourse" about their Black students: They perceived a connection between students' knowledge and behavior, that socioeconomic status was related to students' in-school motivation, and that Black culture and the Black community had an impact on the ways in which students participated in mathematics class. These discourses affected how the two teachers interacted with their students, and much of their classroom practice, which had negative and often explicitly racist overtones. But in addition many researchers note that students also contribute to how they are constructed (and indeed, position themselves), sometimes as a response to how they perceive how teachers see them, and on other occasions based on their own identities and beliefs about schools and mathematics. I describe these issues more fully, as they relate to students, in Chapters 2 and 3.

Many educators have argued that perceptions of teaching are formed before prospective candidates enter preparation programs (Gilbert, 1997; Rushton, 2004). Further, teacher educators concerned with issues of social justice and improvement in urban education believe that teachers' perceptions of urban students, schools, and learning have already been pre-formed before entering an urban classroom (Bell, 2002; Breitborde, 2002; Valli, 1995). As described earlier, popular culture depictions of urban schools and students are often problematic and may reinforce stereotypes of urban students and schools. For a variety of reasons, including the fact that the backgrounds of education students (who are predominantly White) are often incongruent with those of the urban students (who are predominantly of color) they are planning to teach (Bell, 2002; Kea & Bacon, 1999; Valli, 1995), there is a "prevalence of low teacher expectations for ethnic minorities and inner-city students" (Groulx, 2001, p. 61). Teacher education candidates may "make false assumptions about the quality of [urban] schools and of their students and families" (Breitborde, 2002, p. 36). They are particularly reluctant to teach in urban disadvantaged schools (Harris, 2010) and to teach in schools with large populations of African American or Latino/a students (Bakari, 2003). In examining studies of preservice teachers' beliefs on diversity across three time periods (1986–1994, 1995–1999, and 2000–2007), Castro (2010) found common as well as distinct themes across these periods. He notes that researchers discovered across all three time periods that preservice teachers exhibited a "lack of complexity in understanding multicultural issues and contradictory attitudes/perceptions concerning diverse populations and social justice" (p. 200) and, for more recent studies, that teachers' personal backgrounds had an impact on their attitudes, beliefs, and understanding of multicultural concepts. There is also evidence that preservice teachers in recent years have developed an understanding that acknowledging or admitting prejudicial beliefs (particularly with a researcher) is not socially sanctioned in the 21st century. But they may still have these beliefs.

The stereotypes that teachers have about urban schools center around low student ability, student motivation, school safety, and lack of parental involvement (Gilbert, 1997). In a study of preservice teachers, Groulx (2001) found that the teachers "assumed that parent support would be lacking and/or that students would be undisciplined and motivated" (p. 83). Bol and Berry (2005) found that teachers in a sample of National Council of Teachers of Mathematics (NCTM) members believed that a major factor in the achievement gap between White and non-Asian minority students was that "minority students' families did not support teachers or ensure children completed their school work or studying in mathematics" (p. 38). They reported that poor families did not "valu[e] academic achievement or mak[e] education a priority" (p. 38). Some have suggested that the education system (via teacher preparation and also when teachers embark on teaching careers) "socializes novice teachers to emulate the negative attitudes toward teaching African American students that are held by experienced teachers" (Bakari, 2003, p. 643). The truth about Black and Latino/a urban parents' educational values is substantially different, as Hochschild (1995) and others have found. In Chapter 2, I describe the extensive community and parental support for urban, Black, and Latino/a students' education and mathematics performance.

Vera, another graduate student in the same program as Abby, the student whose statement opened this chapter, was herself an alumna of an urban high school. This exchange with the interviewer is revealing in many ways:

INTERVIEWER: Okay. What would you tell your friends about urban high school students?

VERA: Um, just that they're getting worse, that they just don't have the motivation anymore, well, maybe not worse, but, you just have to lower your expectations I guess. . . . Especially in our city, like we have great urban high schools, but it's just sad that students don't care, I think that's the worst thing I would tell people, that they just don't care and you wouldn't understand. . . . Like, they're getting a good education and they just, they won't show up to class. My friends would not believe that if I told them that students just won't show up to class on a regular basis.

According to Vera, the students report that they don't go to class. Her perception that they are unmotivated leads to a statement that one should "lower expectations" of urban high school students. She concedes that some students are willing to hang around after school to get assistance with their work. When the interviewer presses her to differentiate between urban students in general and the students who stay after school for tutoring, the exchange continues:

INTERVIEWER: How about the tutees? Were any of the tutees that you tutored, um, even if they were having problems in mathematics, did any of them seem motivated?

VERA: Um. Motivated in math? Never. [Laughs] But motivated to do well and graduate, I think so. I mean, I guess. I mean they wouldn't be there if not, but it seemed like every one just wanted to get by, you know?

INTERVIEWER: Mm hmm.

VERA: And to some that'd be motivated, but, to me that's not being motivated.

INTERVIEWER: Um, how did you know they just wanted to get by? Like, what did they say, what did they do that made you think that they . . . ?

VERA: Well, the fact that they were there, you know. . . . Sometimes they would see other people leave and I'd go to them and I'd go, "You can leave whenever you want but if you want to work out more problems . . ." and the fact that people said, "No let's do more problems," that shows that they care somewhat, so . . .

INTERVIEWER: Um.

VERA: And they would stay until the end, 'til 5 or even 'til later than 5.

INTERVIEWER: How often did people do that? Did tutees do that?

VERA: 'Til, that stayed for the whole time?

INTERVIEWER: Mm hmm.

VERA: A lot, I'd say. . . . It's just, the thing was sometimes if [the peer tutor] left early with her people, then, maybe some others would leave early, too . . . but I mean it happened pretty much that it went until 5.

INTERVIEWER: So, um, so they would request that they stayed, they would want to stay?

VERA: Yeah, well . . . they weren't like, "Yay, let's stay," but it's like, "We have more problems to do; I really want to make sure I know this," so, I mean that was nice.

Despite Vera's contention that the students "just wanted to get by," her own evidence reveals that many students were committed to staying for the entire duration of the tutoring session. In fact, Vera would approach students and remind them that they "could leave whenever you want," and the students would stay until 5:00, or "even 'til later than 5." Vera, like some of the teachers in Bell (2002), did not "acknowledge [that there were] contradictions within her own rhetoric" about these urban high school students (p. 240).

In the next section (and appearing throughout other chapters in this book), I provide a Lowell High School "snapshot" to describe how the research conducted at Lowell reflects and contextualizes the findings from the research literature. In this snapshot, I describe Lowell teachers' beliefs about their students' mathematics ability and potential, teaching, and mathematics.

AT LOWELL HIGH: TEACHERS' BELIEFS ABOUT THEIR STUDENTS, TEACHING, AND MATHEMATICS

What do teachers in one urban school think about their students, mathematics, and the potential of their students to do mathematics?

In general, teachers at Lowell expressed positive beliefs about teaching and their students. But they did exhibit evidence that they thought that their Lowell students were less capable than the typical U.S. high school student. In a study of the 23 teachers at Lowell in 2004, teachers believed that Lowell students were less prepared for college than the average U.S. high school student and were less likely than the average high school student to succeed in college. More than half of Lowell teachers reported that Lowell either did not prepare students for college work or that they were unsure about whether the school prepared students for college work. Only 4 teachers thought that Lowell students could make good grades in college, if they went.

More specifically, teachers were asked to compare Lowell students with all U.S. high school students on a scale of 1 (more likely than other U.S. students) to 3 (less likely than other U.S. students). The teachers had a somewhat negative view of Lowell students compared with all U.S. high school students. In general, the responses showed that Lowell teachers believed that Lowell students were less likely to do well on the SAT (2.7), do well in math (2.7), and do well in all subjects (2.4). Lowell students were about as likely as the typical U.S. student to go to college (2.1), and to drop out of high school (1.9).

Lowell teachers were also asked about their best students. The typical profile reported by teachers is that their best student is female (17 out of the 23 teachers), earned an A in their class (all 23 teachers), and has friends who are also good students (19 of the 23 teachers chose "about half of them," "most of them," or "all of them"). We asked teachers to further contribute to the profile of their "best student" by telling us if they agreed on a scale of 1 (strongly disagree) to 5 (strongly agree) with certain statements. Teachers strongly agreed (ranging from 4.1 to 4.7) that their best student did well in all subjects, worked well independently, was not disruptive, was well liked, and belonged to a peer group that supported academic achievement. However, they were more neutral about whether the student was the most intelligent person of his or her age whom they knew (3.4) and if the student reminded teachers of themselves when the teachers were high school students (3.2). They had particular beliefs about the kinds of activities that students participated in outside of school, which in some cases did not correspond with what Lowell students reported.

Lowell math teachers expressed commitment to the idea of reform in mathematics teaching and learning, but for many teachers this was not always enacted in their classrooms. In their classroom practice, Lowell teachers tended to adhere to the traditional lecture format, where they explained mathematics concepts, showed examples, and occasionally enlisted student comments. Students then spent the rest of the class period (about 45 minutes) solving problems. The teacher circulated throughout the classroom, assisting where needed. Students most often worked individually, although occasionally they did participate in group work.

The Lowell mathematics teachers ($n = 4$) generally had the same beliefs about students as did nonmathematics teachers, but were also asked questions specific to mathematics. On a scale of 1 (strongly disagree) to 6 (strongly agree), mathematics teachers reported that motivation (5), practice (5.3), and exposure to challenging problems (4.5) most influenced student improvement in mathematics. Teachers also reported that teachers' belief in students' ability (4.5) and the socioeconomic status of students' families (4) also had a strong influence on student improvement.

Math teachers were asked about the characteristics of the best mathematics students, using a scale of 1 to 6. They described them as achieving high mathematics scores on standardized tests (5.5), typically doing well in school overall (4.5), and working hard when solving mathematics problems (5). They were more neutral about whether or not the best students had to work hard to get good grades in mathematics (3.5), had parents who were active and involved (3.5), didn't talk a lot in class (3.8), and worked alone most of the time (3.5). These suggest that Lowell math teachers thought in traditional ways about measures of high performance in mathematics. They may believe that the best students are "naturally good" in mathematics, for example.

Math teachers were also asked about what they felt would most influence student improvement in mathematics. They most strongly agreed (4.5 and above) in the order presented here that practicing procedures, practicing problem-solving techniques, student motivation to improve, exposing students to challenging problems, and the teacher's beliefs that the student could be a good mathematics student were strong contributors to student mathematics improvement. Teachers felt that students' socioeconomic status (4), whether or not the student was disruptive in class (4), and memorizing skills (3.8) also influenced students' improvement in mathematics.

On a series of questions adapted from Raymond (1997), math teachers responded along a continuum denoting their agreement with characterizations of mathematics, mathematics learning, and mathematics teaching. They strongly felt that mathematics was a dynamic, expanding body of knowledge; were more likely to characterize math as surprising than predictable; were more likely to say it was difficult most of the time than to say it was easy most of the time; and were more likely to feel that doing mathematics involved working alone to solve problems rather than involved working collaboratively. They felt strongly that learning mathematics required mostly practice (rather than intuition), independent work (rather than group work), trying hard (rather than being good at math), and understanding (rather than memorizing). Finally, they felt strongly that good mathematics teaching depended on student effort (rather than teacher effort), flexible lessons (rather than explicit planning), and helping students to self-assess and correct their own mistakes (rather than showing students their mistakes and demonstrating how to solve a problem correctly). They were more likely to agree that good mathematics teaching depended on helping students to see mathematics as useful (rather than helping students to like mathematics).

What is interesting about these responses is that the first category of statements (about *mathematics*) allows teachers to share their philosophies and, perhaps, their affective beliefs about mathematics. Their responses echo findings that math teachers adhere to "reform" philosophies about mathematics. But in the third category, about *mathematics teaching*, the practical takes precedence over the idealized. Good mathematics teaching depends a lot on students' actions in the classroom, and these teachers (given their belief that it is more important to help students see mathematics as useful than enjoyable) may be hammering home the perspective that math is best used for utilitarian or instrumental purposes. Finally, despite their strong agreement that good teachers helped students to self-assess and correct their own mistakes, this was very rarely the case in Lowell classrooms—and most American ones, for that matter (Stigler & Hiebert, 1999).

STUDENTS' OTHER MATH OUTCOMES: ATTITUDES AND PARTICIPATION

Earlier in this chapter, as part of a discussion about the achievement gap, I discussed national data on students' mathematics achievement. But mathematics outcomes are not limited to test scores. Other important factors—such as students' dispositions toward mathematics and their rates of participation in mathematics—contribute to students' test scores, but also to the development of students' mathematics identities. Eventually, how students feel, perform, and participate in mathematics affects if and how they see themselves as mathematically competent.

Attitudes

In a previous section I described students' achievement levels in mathematics based on standardized assessments like the NAEP. But how do students, especially urban, Black, and Latino/a students, feel about mathematics? Overall, national data show that students tend to dislike mathematics, although they feel that they perform at average or above-average levels in mathematics classes in school. (A recent international assessment, the Program for International Student Assessment [PISA], found that U.S. students were among the lowest achieving but had some of the highest confidence levels in their mathematics ability compared with their international counterparts!). The longer students are in school, the less interested they are in mathematics—students are much more positive about mathematics in elementary school than they are in high school, for example. This pattern is true across all demographic groups as measured by the NCES. For example, in 2000, 70% of 4th graders agreed that they "like mathematics," compared with 54% of 8th graders and 47% of 12th graders. Older students are also less likely to agree with the statement "All students can do well in mathematics if they try" (88% at grade 4, 71% at grade 8, and 42% at grade 12).

When data are disaggregated by race/ethnicity and, further, by gender, however, an even more complex picture emerges. National data suggest that Black students especially have very positive attitudes toward school, learning, and mathematics, often exceeding those of their ethnic counterparts. In 2000, for example, higher percentages of Black students at most grade levels of the NAEP (4th, 8th, and 12th grade) agreed with the statements "I like mathematics" and "All students can do well in mathematics if they try" (90% at grade 4) than those of any other ethnic group. Black and Latino/a students, who constitute the majority of urban students, are among the most interested in mathematics (Catsambis, 1994; Strutchens et al., 2004). But these are also the students who perform the least well, on average, on standardized assessments. Roselyn Mickelson (1990) refers to this as the "attitude achievement paradox"—that despite students' positive attitudes, their achievement levels don't match their enthusiasm. In addition, despite students' positive attitudes, teachers of students in urban schools often have negative perceptions of their abilities and potential to do rigorous mathematics and assume they are disinterested.

In terms of students' reported confidence and understanding, however, the picture is somewhat different. At all grade levels, a majority of students report that they are good in mathematics and that they understand most of what goes on in mathematics class. However, in 4th grade, 73% of Black students and 69% of Latino/a students agree with the statement "I understand most of what goes on in mathematics class" compared with 80% of White students. In 12th grade, 54% of White students agree that they are good in mathematics, compared with 50% of Black students and 45% of Latino/a students. Students also have intriguing beliefs about the nature of mathematics. A higher proportion of Black and Latino/a students than White students believe that "learning mathematics is mostly memorizing facts" at all grade levels. While 25% of Black and 24% of Latino/a 4th graders believe that "there is only one correct way to solve a mathematics problem," only 13% of White 4th graders do. By 12th grade, these differences have dissipated—8% of Latino/as, compared with 5% of Blacks and 5% of Whites, believe this.

In addition, there are intriguing trends related to gender within racial/ethnic groups. Regarding achievement, there is evidence that achievement gaps in mathematics performance, where they exist, favor boys only for White students. The only statistically significant gap between males and females on the 2000 NAEP existed in 4th grade for White students, and was a gap of 5 points—not of practical significance. Further, this difference is largely due to differences in achievement among White students at high socioeconomic status (SES) levels. At grades 4, 8, and 12, there is a statistically significant gap between males and females, favoring males, for White students of higher SES. (Interestingly, these data disaggregated by race, gender, and SES show that the achievement at the 4th-grade level for Black and Hispanic students of higher SES is comparable with that of White students of lower SES.)

The research on girls' attitudes toward mathematics suggests that on average, girls tend to view mathematics as a male subject that will not be useful to them (American Association of University Women, 1992; Fennema & Leder, 1990) and have less positive attitudes toward math and math class than do boys (Catsambis, 1994). Strutchens's analyses of 2000 NAEP data bear this out: In 2000, at all grade levels and on all three statements "I like mathematics," "I am good at mathematics," and "I understand most of what goes on in mathematics class" (except for "I understand . . . " at the 4th-grade level, where there was no gender difference), the percentage of male students agreeing with those statements is higher than that of female students, and that difference was statistically significant. In 12th grade, for example, 47% of female students agreed that they were good at mathematics, compared with 58% of male students. In addition, 12th-grade girls were less likely to agree than 12th-grade boys that "all students can do well in mathematics if they try."

However, there is evidence that there are important intersections between ethnicity and gender in terms of attitudes toward math. Catsambis (1994) found that although Black students, regardless of gender, reported equally positive attitudes toward mathematics, by 10th grade girls of all ethnic groups were less likely to agree than their male counterparts that math was one of their best subjects, although Black girls still had more positive attitudes than their White counterparts (Riegle-Crumb, 2006). These findings suggest several important phenomena: first, that girls in 8th grade may be affected by a perception that mathematics is a male domain; second, in the early years of high school White and Latina girls may be more affected than Black girls by gender stereotyping; and third, and perhaps most troubling, that by 10th grade, girls, regardless of ethnic background, either legitimately feel that math is not their best subject or are hesitant to report that they are good in mathematics. Some researchers (e.g., Campbell, 1995; Fennema & Leder, 1990) posit that these findings reflect differential expectations of math ability for girls and boys by teachers and parents.

Participation: Course-Taking and Mathematics Behaviors

Course-taking differences, as described earlier, are critical to understand because evidence shows that students who take more, and more advanced, courses score higher on assessments (NCES, 2009). In addition, taking courses in the advanced mathematics pipeline is an important signal to colleges and universities and a marker of postsecondary access and attainment.

But there is evidence that profound gender differences have been masked in previous research. While a significant body of literature about gender differences in mathematics participation exists, this work (e.g., Fennema & Leder, 1990) generally has not considered gender differences *within* ethnic groups. The general conclusion is that the gender gap in mathematics course-taking has generally favored males; but Catsambis (1994) suggested that the established gender effect may be reversed for some ethnic groups; for example, Walker (2001) found that Black male students are less likely to persist in advanced mathematics course than Black female students in high school. Among 10th graders, White and Black girls (but not Latino/a girls) are more likely than their male counterparts to say that they are enrolled in an academic curriculum and to have taken geometry (Catsambis, 1994). In addition, there is substantial evidence that even when Black and Latino/a students "start out" at the same critical entry point, Algebra I, at the same time as their White counterparts, racial/ethnic and gender differences emerge in more advanced courses later in the pipeline (Riegle-Crumb, 2006; Walker, 2001). While for some students this is due to lack of availability of advanced courses in their schools or other "inequality that exists in high schools" (Riegle-Crumb, 2006, p. 117) a key finding in Riegle-Crumb's study was that African American and Latino boys do not benefit from taking Algebra I "on-time" (Walker, 2001), as do their Black and Latina female counterparts and their fellow White students. The fact that Black and Latino boys, in particular, are significantly underrepresented in advanced mathematics courses has critical consequences for their test performance and postsecondary access and career success.

AT LOWELL HIGH: STUDENTS' PERFORMANCE, ATTITUDES, AND BEHAVIORS

Mathematics Achievement Overview

On average, Lowell students ($n = 154$) reported on a questionnaire that their current grade in their mathematics class is a C, with 15% of students reporting a current grade of A, 27.9% reporting a current grade of B, 31.6% reporting a current grade of C, and 29.4% of students reporting a current grade of D or F. Generally, male students reported higher grades than females, roughly a B- compared with girls' average grade of C. Black girls and boys reported similar mathematics grades of C, but Latino boys reported slightly higher grades than Latina girls (average grade of B- for boys, C+ for girls). This contradicts national data that suggest that girls get better grades in math than boys, or might reflect that boys may be more likely than girls to inflate their grades when self-reporting.

Mathematics Attitudes, Behaviors, and Future Plans

We asked students several questions about their mathematics attitudes and behaviors using a Likert scale (1 = strongly disagree, 3 = neutral, 5 = strongly agree). Generally, students report neutral feelings about mathematics. Like high school students nationally, they do not seem to particularly enjoy it (2.88), report that they are not very good at it (2.70), and are not very apt to report that it is their favorite subject (2.24). They do recognize that mathematics is important for their future—they agree that it is useful in everyday problems (3.48), that they will use it in many ways as an adult (3.82), and that it is important to know math to get a good job (3.96) and to get into college (4.31). In terms of their attitudes about their own math work specifically, they agree that they usually understand most of what is done in math class (3.46). However, they do not spend a lot of time studying for math tests (2.32) and they are neutral about time spent doing math homework (3.00). Despite students' recognition that math is important, they are less likely to agree that they do better in math than most people (2.63) and that people who share their race/ethnicity do well in math (2.63). These last two findings are important to note because if students do not feel particularly efficacious in math, they may not put in the amount of effort required to do well in it. If teachers exhibit behaviors that suggest that they think mathematics is "something you either get or you don't," students may think that their lack of success in math is due to an inherent lack of ability, rather than lack of effort.

We also discovered some intriguing differences between male and female students in their mathematics beliefs and behaviors. Although boys report higher grades overall, girls are more comfortable than boys presenting math work in front of the class (48% versus 30%). Twice as many boys in the sample as girls reported that they enjoyed mathematics. Three times as many boys as girls (48.9% vs. 16%) reported that they would consider majoring in math or a mathematically related field in college. While 23.3% of girls stated that they strongly agreed that they were not very good in math, only 9.4% of boys agreed with this statement. Yet only 28% of boys reported that their friends would admire them for doing well in math, compared with about 74% of girls. This will be discussed further in Chapter 2.

In line with their slightly higher reported grades in math, male students more strongly agree that they enjoy math than female students (3.65 vs. 2.56, $t = 4.918$, $p < .01$). Male students also more strongly agree than do female students that their current math teachers would recommend them for an honors or advanced math course, and that their teachers think they are good in math. However, Black girls are more likely to report that their current teacher thinks they are good in math than are Latina girls, and Black boys are more likely to report that their current teacher cares about how well they do in class than are Latino boys.

At Lowell, there are fewer attitudinal differences between Black boys and girls than there are between Latino boys and Latina girls. Both sets of boys are more likely to report that their teacher would recommend them for an honors or advanced class than girls are. While Black boys are more confident in their mathematics abilities than are girls, and are slightly more likely to agree with the statement that boys are better than girls in mathematics, differences between Latino boys and Latina girls are more extensive. Latino boys report enjoying mathematics more and are more likely to agree that they usually understand most of what is happening in math class, that it is important to know math to get into college, that they do better in math than most people, and that math is their very favorite subject.

The gender differences in attitudes and mathematics behaviors within ethnic groups reported by Lowell students, which are all statistically significant, are intriguing and may be related to differences in students' peer communities. The next chapter explores students' peer communities as well as other kinds of networks that support their mathematics development.

CHAPTER 2

Understanding Students' Communities and How They Support Mathematics Engagement and Learning

As described in Chapter 1, urban students, and in particular Black and Latino/a students attending urban schools, are often characterized as being disinterested in school and in mathematics.[1] These characterizations, popularized by the press and popular media, promulgate stereotypes of Black and Latino/a youth. Further, when these young people's peer groups are depicted, the groups are most often shown as negative influences: more interested in extracurricular, noneducational, and illegal activities than in school-based activities. Their families are depicted as disinterested in their students' success and as incapable of providing assistance and information that helps students excel in school. In many ways, we think we know a great deal about urban students' academic communities: We assume that they do not have one, that high-achieving urban students are in conflict with their peers, and that they have to compensate for community and parent "deficits." Unfortunately, many school adults (teachers, counselors, administrators) share these beliefs (as described in Chapter 1) and may construct opportunities to learn mathematics accordingly, as I will describe in Chapter 3.

In this chapter, I examine how urban students characterize their peer and family networks and how these support or obstruct their mathematics success. I draw upon existing research as well as the work at Lowell High to discuss these networks. But first I address another common myth about Black and Latino/a communities in general: that they do not support education or mathematics.

BLACK AND LATINO/A COMMUNITY SUPPORT FOR EDUCATION AND MATHEMATICS LEARNING

Black and Latino/a community involvement in education is both extensive and long ranging: in recent years, historical and contemporary community engagement in education in both the African American (Anderson, 1988; Gunn

Morris & Morris, 2000; Jordan-Irvine, 2000; Mirel, 1999; Perry, Steele, & Hilliard, 2003; Siddle-Walker, 1996) and Latino/a (Auerbach, 2002; Conchas, 2001; Moreno, 1999; Stanton-Salazar, 2001; Valenzuela, 1999) communities has been described. It is important to understand that there is a legacy of achievement orientation in underserved communities that may be useful to our understanding of student achievement today. Recent work (Cooper et al., 2002; Guajardo & Guajardo, 2004; Moreno, 1999; Perry et al., 2003) highlights the intellectual heritage of people of color and the role of their community traditions (e.g., historically Black colleges and universities, churches and other religious organizations, social clubs, and political organizations) in challenging limiting and limited education and developing academically successful individuals. In particular, the academic success of many African American students during the era of segregation despite odds imposed by societal mores and governmental agencies, both within segregated schools and in nominally desegregated schools, was largely due to an ethos facilitated by supportive social networks consisting of relatives, community members, and others who may or may not have had a close familial relationship with individual students (Gunn Morris & Morris, 2000; Perry et al., 2003; Siddle-Walker, 1996). These social networks formed an academic community that fostered academic success.

Historical studies of Black and Latino/a educational communities (Anderson, 1988; Cooper et al., 2002; Franklin, 1990; Guajardo & Guajardo, 2004; Gunn Morris & Morris, 2000; Moreno, 1999; Siddle-Walker, 1996) reveal that, as communities of learners, Black and Latino/a students were driven to excel by each other as well as by older students (including their own siblings) and adults (both parents and supportive teachers and school administrators). Certainly in contemporary research focusing on adolescent students, there is a fundamental premise that students' peer groups may be critical in the development of an identity that supports academic success (e.g., Azmitia & Cooper, 2001; Datnow & Cooper, 1997; Howard, 2003; Steinberg, Dornbusch, & Brown, 1992; Yonezawa, Wells, & Serna, 2002).

This research is in marked contrast to the prevailing view of many that Black and Latino/a parents are disinterested in students' education. The current discourse about the academic achievement of students of color has largely neglected the important context of their community's support for education and positive peer influences (for notable exceptions, see Perry et al., 2003, and others). Further, throughout the literature, there is a supposition that these networks that historically supported academic achievement have eroded. Now, in an era in which most urban schools are segregated nearly 60 years after the *Brown* decision rendered school segregation unconstitutional (Orfield & Yun, 1999), a seemingly common (and largely anecdotal) perception is that there is less support for education, academic achievement, and academic activities in general from members of these communities.

AT LOWELL HIGH: FAMILY SUPPORT FOR MATHEMATICS LEARNING

Much of the literature on high student achievement for underserved communities focuses on parents' contributions and expectations, and perspectives from a sample of Lowell High's high-achieving students, who were interviewed in 2004, reveal substantial family and community support for education and mathematics. Supporting this, all the students ($n = 21$) mentioned that their parents were integral to their success in some way. They spoke about their parents' expectations that they would earn good grades, and also, in several cases, that their parents' approval was important to them and, further, that they did not want to disappoint their parents:

> My mom would be the first person [who influences my success]. Her expectations of me are very high, I can do it, but if I don't put anything into it then she'll probably get mad 'cause she knows that I can do it. (Yvette)

> My parents, they encourage me to do well in math. . . . I don't know, I want to make my parents proud. (Lourdes)

> Like, my mom, she likes when I get good grades in math. (Lamont)

> I've been doing well since I was little, so I always try to keep it up. . . . So then, that way, I won't disappoint my mom. (Kayla)

> My father is big on school and stuff like that. So he's the one who pushes me the most. (Adriana)

> My pops, he was pretty smart in school, or whatever. And then, you know, like he encourages me to do well, plus if I don't do well I get punishment. (Ian)

In addition to parents providing high expectations and consequences for low achievement, students also reported the ways in which their parents, particularly when the students were younger, helped them do well in mathematics. Isabel noted that a major reason for her success in math was her mother. "My mom likes math also, and she used to help me when I was young with my homework. And she gave me problems I hadn't seen at school." Ellen mentioned that her father did mathematics with her when she was younger:

> At a real young age he would teach me how to count. You know, like fun things for kids: count the cookies, or count the eggs, or stuff like that. So he would do things with me like that, and I really caught on fast, so I knew my times tables by the 2nd grade. All of them. And he used to test me like,

"2 times 2," and I would say, "4," and he would go, "4 times 4," and I would go, "16," and we'd go on like that.

Esteban mentioned his parents' high expectations for grades, but when asked if he ever talked to his parents about what he learned in math class, he laughed, saying, "No, my mom, she'd just get a headache." Several students also mentioned that although their parents did not explicitly help them with their current mathematics work, they supported them in other ways, in particular, by encouraging them to do their homework and schoolwork, and by telling them to "take advantage" of opportunities students have but that parents may not have had:

My mom, she never went to school, but she loves math. She's like you should do well, and math helps you. My mom is like a role model, but she never helps me with math because she didn't go to school a lot. (Ana)

I know that my pops messed up in school, you know, he used to hang out and all that and I learned from his mistakes. . . . He told me about him messing up. Even though he got a GED he had like 80s or 90s. He was saying that he was good, but you know, [you have to] take advantage. (Ian)

Other students reported that other family adults, in addition to parents, contributed to their success in mathematics. For example, Lamont reported, "In 5th grade, I was in a really gifted class and the math was a lot harder than I usually had so it would be my uncle who helped me out a lot."

It should be noted that familial influences included those of "near peers," not just adults. Students' siblings and cousins also served as encouraging models for students. Lena wrote on her map of influences (see Appendix B for a sample map) about her cousin, Chris: "My role model. I look up to him. Math is his favorite subject. He also pushes me."

Datnow and Cooper (1997) showed that high-achieving independent school students served as models of academic behavior for their younger peers. As they did for some of their friends, Lowell students also served as role models and tutors for their younger siblings:

My brothers and sisters look up to me so if I do it well then they won't see it as so hard, and then I can help them. . . . . I could help everyone else. (Yvette)

Other Lowell students benefited from this model, in that they were helped by or received advice from older siblings and cousins (notably, three students reported that they had siblings and cousins also attending Lowell).

My brother and my family's in the same class as me, if you look outside right there he's tops on there all the time [on Lowell's honor roll]. . . . We're all good in math, we always have been. (John)

My sister is also good in math and I want to be like her. (Lourdes)

[My older brother] helps me with homework if I don't understand. (Elizabeth)

I remember like in first grade I had problems with numbers so [my big sister] started helping me with math. . . . She is 18 and in college. (Gabriela)

While most students reported that their siblings were either positive influences on their mathematics achievement or were younger peers to serve as models for, a few saw their siblings as people who were not to be emulated. Kayla reported that she did not "want to be compared with them [my siblings]; they messed up." This served to drive Kayla to achieve in school so that her mother would be proud of her.

These students want to emulate people in their lives whom they see as strong, smart, and supportive, even if those people did not graduate from high school. Others (particularly school adults) might not understand why students would choose to emulate these people. But these students do not necessarily see as negative the fact that some of their parents did not receive much formal schooling. Even though these parents may not be able to help them with specific mathematics work, their encouragement, expectations, and "lost dreams" are powerful motivators for these students. Students reported that they do not want to repeat the missed opportunities of their parents or the mistakes of parents or siblings. In particular, this is true of Lowell's Latino/a students, similar to Auerbach's (2002) findings that Latino parents provide "counterexamples" for their children in terms of the types of futures they want their children to avoid (Romo & Falbo, 1996). But in addition to support from their families, students report support from their peer groups and describe vividly how students support themselves and each other in mathematics.

PEER INFLUENCES AND ACADEMIC ACHIEVEMENT

Although contemporary accounts of the education of African Americans, specifically, before and after the civil rights movement, have begun to shed light on these positive communities, they have not explicitly focused on the positive role of peers in Black academic achievement. For example, Perry et al.'s (2003) narrative analysis comprehensively describes the supportive roles of family and community in fostering African American success, but it does not explicitly address the positive role of peers in facilitating academic success, although this is implicit in some of the narratives included. There are critical differences between what we know to be the historic roles of communities in developing and supporting the academic success of Latino/a and Black students and the ways in which we currently view these students' peer groups and their impact on intellectual identities.

Academically successful African Americans and Latino/as have acknowledged the implicit and explicit support of their immediate peer groups, "near peers" (those who may not necessarily have been close to them in age but were still of their "generation"), and mentors. In a study with African American mathematicians (Walker, 2010, 2011), many describe the influences of peers on their mathematics success, from elementary and secondary school through graduate school and their careers. One African American mathematician describes a high school classmate exhorting him to do well, to prove that African Americans could do well in math:

> He said, "Look, you have a responsibility." I still remember to this day, he says, "You're better than any of us in terms of doing this stuff." And he says, "You're probably better that most of the White students." He says, "You got to stay number one, and you also have an obligation to help, you know, to tutor and stuff like that." So you know, anybody that was kind of interested I would help them. Not because of him, I would have done that anyway. But I did feel this obligation because he would monitor what I was doing. . . . He told me that that was my obligation, and I kind of believed him. There were times when I kind of didn't feel like studying, I would kind of like hear his voice. Which was really kind of interesting to me. And I wouldn't remember his name or his face if he would walk up to me now at all. In a way I would like to thank him.

In describing his counselor's and other administrators' attempts to keep him and some of his high school classmates out of advanced mathematics and science classes, Tate (1994) wrote about the students' collective efforts to fight these attempts and his subsequent efforts to interest classmates, friends, and family members in doing mathematics. Flores-Gonzalez (1999) reported that the Latino/a students in her study formed a visible high school peer group that endorsed high achievement and promoted academic behaviors. Davis, Jenkins, and Hunt (2002) have written powerfully about their experiences as young African American men working together to become doctors of medicine and dentistry. Their self-supporting peer network endorsed important academic behaviors, including their pushing one another to continue to take advanced mathematics and science courses, as well as persisting in college. Their book illustrates that they benefited from one another's encouragement.

A substantial body of research reveals that adolescents of all ethnicities seem to lose interest in education and mathematics as they progress through school (Osborne, 1997; Steinberg et al., 1992; Strutchens et al., 2004). The later elementary school and middle school years are the times when the values of adolescents' peer groups begin to supersede the values of their parents and families, and according to many educators, the peer group focus shifts from academic to nonacademic activities and has significant influence on students' beliefs and behaviors about school and academic achievement (Steinberg, Brown, & Dornbusch, 1996). During adolescence, parents may express concern that the academic influences of their students' peers outweigh the parents' own positive academic influences (Auerbach, 2002; Azmitia & Cooper, 2001; Stanton-Salazar, 2001; Steinberg et al., 1992, 1996). Further, ethnic minority students say that their peer group often presents obstacles to achievement (Azmitia & Cooper, 2001; Cammarota, 2004; Martin, 2000; Ogbu, 2003).

Some students of color may have to respond to challenges from peers (Azmitia & Cooper, 2001; Conchas, 2001; Ogbu, 2003) who question their allegiance to their ethnic group. If the peers' perception is that certain students are not adhering to the peer group's norms around achievement, and these norms are not counteracted by school adults (Polite, 1994), the students may find themselves constantly negotiating between their friends' perceptions of them and their own academic selves. The dominant peer group norms, especially if they do not support academic behaviors, can exacerbate a climate of underachievement, particularly for African American and Latino/a students.

One widely held theory regarding African American and Latino/a underachievement and the role of peer influences on achievement comes from the work of Fordham (1988), Fordham and Ogbu (1986), and Ogbu (1987). This theory suggests that "involuntary minorities" (Ogbu, 1987), including African Americans and Latino/as, eschew schooling because they do not see it as a legitimate enterprise in improving the odds of social mobility for people who share their ethnic heritage. In Fordham's (1988) ethnographic work, adolescents reported that students who did well academically were seen as "acting White." Even though some have challenged Fordham's and Ogbu's theories (Ainsworth-Darnell & Downey, 1998; Cook & Ludwig, 1998; D. H. Ford, Harris, Webb, & Jones, 1994), noting that they do not provide a framework for understanding ethnic minority students' success (Conchas, 2001), it is important to examine them because they have had a significant and pervasive impact on how educators view African American and Latino/a students and their response to schooling. In particular, the "acting White" hypothesis has gained substantial currency in the media, often oversimplified (Horvat & Lewis, 2003) and mentioned in multiple news outlets as an explanation for ethnic minority underachievement.

In support of Fordham and Ogbu's (1986) theory, there is qualitative evidence that ethnic minority students may avoid taking certain advanced courses for social reasons—they want to attend classes with their friends, not be the sole ethnic minority student in certain classes, and not "stand out" academically (Polite, 1994; Walker & McCoy, 1997). In particular, academically successful students may attempt to minimize their academic success in order to be accepted by their peers (Cooper et al., 2002; Ford, 1996; Fordham & Ogbu, 1986; Walker & McCoy, 1997), who may not value academic success or consider it to be the domain of students from other ethnic groups. Using data from the National Education Longitudinal Study of 1988 (NELS 88), Cook and Ludwig (1998) found that there was no difference in social penalty incurred for success for Black and White students; that is, high-achieving Black and White students were "no more likely to be unpopular than other students" (p. 391). But Fordham's (1988) work suggests that Black students may adopt a stance of "racelessness," in which individuals "assimilate into the dominant group by de-emphasizing characteristics that might identify them as members of the subordinate group" (Tatum, 1997, p. 63). Several popular and empirical works paint the picture that academically successful Black students, more so than their White counterparts, may often have a tough choice: do well in school and be a loner or conform to group norms about academic performance and have a social life (Bempechat, 1998; Steinberg et al., 1996; Suskind, 1998).

Ogbu and Davis's (2003) work examining African American students' academic achievement in Shaker Heights, Ohio, suggests that Black students' academic behaviors reflect their beliefs that performing well lessens their acceptance by their Black peers. Some have suggested that "acting White" is more of an issue within predominantly White settings than within predominantly Black settings.

Many researchers have argued that a different social penalty may be paid by underrepresented students of color versus White students. We know very little about these particular mechanisms, and further, we know little about the peer groups of Black and Latino/a students who are academically successful, beyond notions that such students are "loners" and socially inept. Nor do we know how strong the "codes" about academic achievement are in these students' peer groups, and how they affect student achievement (Yonezawa et al., 2002). Undoubtedly, there are many students who do well academically and maintain connections to their ethnic group (Conchas, 2001; Datnow & Cooper, 1997; Flores-Gonzalez, 1999), as Horvat and Lewis (2003) discovered. They found significant variation within students' peer groups, and describe vividly the ability of students to maintain what they consider an authentic Black identity while navigating multiple peer groups and those groups' perceptions of their success. Indeed, it is possible that peer support for Black and Latino/a students' academic achievement is not as dichotomous (Ogbu, 2003) as much of the research suggests. And it is critically important to note that some peers may be more influential in terms of students' academic beliefs and behaviors than others (Harris, 2010).

Fordham and Ogbu's (1986), Ogbu's (1986, 1987), and Ogbu and Davis's (2003) work has contributed to the notion that Black students in particular may experience social sanctions for high academic achievement. But an illustrative quotation from Ogbu and Davis's (2003) work reveals that Black high school students think broadly and with complexity about peer academic support (or nonsupport):

Another student, however, reminded the group [of researchers and high school students participating in a focus group] that "The picture is not so dichotomous." There were some positive peer pressures to do well in school among Black students. Examples of positive peer pressures described for us at interviews included the following:

Anthrop: What about peer pressure, how important is that to how well a student is doing, or how poorly a student is doing?

Student: Um, it depends upon how the . . . person was brought up because we all have different um, levels of . . . vulnerability to peer pressure . . . Um, some people just shrug it off and some people are really affected by it.

Anthrop: Mh-hmm.

Student: But . . . I do think it in some way it does have, like you know, the most minute effect on us . . . So um, if, you know, you hang around people that are doing well in school, you're gonna have a tendency to do well in school. If you hang around people that are the opposite, you might do bad. So, it does have some sort of effect on us. There's pressure that we can apply upon other people. Then, if you're in the case where you're doing well, and other people aren't doing so well, um, you know, you might [influence them] to bring their grades up. So it does work both ways. (p. 192)

The prevalent interpretations may lead to an oversimplified characterization that assumes that all Latino/a and Black students respond to schooling, whether or not they consider it oppressive, in the same ways: by either overtly resisting schooling or by not trying hard. This trope is becoming an increasingly common one to explain Black and Latino males' performance in school. On multiple outcomes—high school graduation rates, grades and GPA, and college degree attainment—Black and Latina females have better outcomes, on average, than those of their male counterparts (Riegle-Crumb, 2006). In fact, in personal communication with Horvat and Lewis (2003), Ogbu noted that "he and other ethnographers have found that young Black women are less affected by the burden of acting white" (p. 268). There is research evidence that male students may be less likely than female students to benefit from a peer culture that supports academic success. For Black boys in particular, "many academic problems surface due to African American males' rejection of academic traits as being European or feminine. African American males are more likely to deny and devalue academic interests to avoid the ridicule and shame that accompany success" (Corbin & Pruitt, 1997, p. 74). But the suggestion that Black and Latino young men unilaterally have an oppositional stance to schooling obscures the academic success of young men who are successful (Berry, 2008; Moore, 2006; Stinson, 2006; Tate, 1994) and reduces the real challenges they face in schools and classrooms, from the early elementary years throughout their educational careers, to something that is completely under their control, when it is clear that there are school processes and mechanisms that obstruct success. Osborne (1997) described a troubling finding from his study of self-esteem, academic performance, and ethnicity using NELS 88 data: that unlike other ethnic groups, by the time that Black male students were in 12th grade, their academic performance and self-esteem were completely uncorrelated. For all other groups, academic performance and self-esteem were positively correlated. Cooper (2003), A. A. Ferguson (2001), Polite and Davis (1999), Noguera (2008), Stinson (2006), and others suggest that there are numerous signals and messages that Black and Latino boys receive about their value as students both within and outside of schools that do not support their academic development. Cammarota (2004) specifically points out that "media and public policy discourses advance both implicit and explicit assumptions that urban males pose significant threats to public safety" (p. 57) and that these assumptions are reflected in how Black and Latino boys are "adultified" (A. A. Ferguson, 2001) and punished more harshly than others in schools (Noguera, 2008).

But it is certainly possible that some Latino/a and Black students respond to school in ways that support academic achievement. For example, some researchers have found that because of the pervasive intellectual stereotypes about African Americans and Latino/as, students may engage in academic activities as a "mission" to prove that they are not intellectually inferior (D. H. Ford et al., 1994; Perry et al., 2003; Valenzuela, 1999; Yonezawa et al., 2002) and that "being Black and being smart are not incongruent" (Horvat & Lewis, 2003, p. 276). Datnow and Cooper (1997) discovered that Black students attending predominantly White independent schools relied on formal and informal peer networks for various kinds of academic support. This support ranged from the "modeling" of academic behaviors (by a student who studied in the library rather than spending free time playing in a basketball game) to the explicit help and tutoring of younger students by older students (Datnow & Cooper, 1997). Cooper et al. (2002) reported that peers' positive influences contributed to Latino/a student success in a college-outreach program. Thus, it is entirely possible, and indeed probable, that successful African American and Latino/a students draw different types of support from peers who may or may not share their academic success, in ways that we still do not fully understand (Azmitia & Cooper, 2001; Flores-Gonzalez, 1999). In the Horvat and Lewis (2003) study, they found that academically successful Black students "camouflaged" (Fordham & Ogbu, 1986) their success when engaging with unsupportive peers, "code-switched" when speaking with different audiences, and participated in multiple social networks. But these behaviors were not necessarily due to a fear of ostracism, but rather were engaged in for multiple reasons—to not hurt friends' feelings, for example. Like Walker (2006), Horvat and Lewis (2003) found that students were adept at navigating multiple social worlds while performing well academically.

It is clear that successful urban high school students benefit from attentive and interested parents and supportive peers who express in various ways that they share students' commitment to and interest in education. (In a later chapter I will discuss how committed and caring teachers also constitute an important part of urban students' academic communities around mathematics). Evidence suggests that Black and Latino/a support for education remains strong (Azmitia & Cooper, 2001; Hochschild, 1995; Solorzano, 1992; Stanton-Salazar, 2001). Yet students' peer groups may provide support that goes unnoticed by parents or school adults, support that can be useful in improving achievement for underserved students. It is this limited understanding of urban high school students that contributes to school practices that may negate or undervalue positive academic behaviors and attitudes fostered by students' parents and peers.

Because urban students' peer groups have been maligned in the popular press and viewed in limited ways by many educators and education researchers, the initial impetus for my work at Lowell High was to document the ways in which students' peer groups might or might not support mathematics learning. What I found was revealing and illuminating.

AT LOWELL HIGH: STUDENTS' PEER GROUPS AND THEIR SUPPORT FOR MATHEMATICS

Unlike the students in Flores-Gonzalez's study (1999), the high-achieving students at Lowell did not perceive themselves to be officially labeled by other students as an identifiable "high-achieving" group, although individually they were known for doing well in mathematics. As with the students in Horvat and Lewis's (2003) study, their friendships and social networks in school encompassed students who were successful in school, as well as those who were unsuccessful.

For example, Linus described his relationship with his friend Andrew and how other friends commented on their mathematics prowess:

> Like me and my friend, Andrew, we are the good ones in math. And people always ask us how do we understand that, and how do we know it before the teacher teaches it?. . . And like my friends ask me, "How do you do that? How do you understand that if he didn't even teach it yet?"

Later, Linus pointed out that he and Andrew work together on mathematics problems when they don't understand something. But Linus also added that he is pushed by family members outside of school:

> I am the youngest of all of the cousins, so they push me. . . . People always push me, like, a lot of it is in math, and my brother is like, "Come on, you've got to compete with me, you've got to be up there with me."

Linus's older sister also had high expectations of him:

> My sister, she overrates me. Like if it's not 85 or higher [his grade on a test], 90 or higher, she goes, "Why get an 85? You only know 85% of this stuff?" Since she doesn't have a good math background, she wants to see me do good in math. She doesn't want me to mess up in math like she did. She used to have a lot of problems. She used to pass the class, but she would struggle.

Linus's relationship with a friend who does not attend Lowell indicated that his academic community encompassed people who were outside of his immediate sphere:

Like, I have a friend who moved to Florida, my best friend, and whenever we talk he's like, "You still the king of math?" and I'm like, "Yeah." So he knows that I'm good at math and that he's good at math, too. We went to junior high together and the teacher used to separate me and him from the rest because people used to try to copy off our tests. So they used to sit me in one corner with him.

Varied Peer Responses to High Mathematics Achievement

High-achieving students experienced several types of support from their friends and peers, inside and outside of Lowell. In addition, students reported a mix of responses to their success from peers: Some of their peers were indifferent to their success, and some congratulated the students on their mathematics prowess, "good-naturedly" referring to the student as a "genius" or a "nerd." Students reported little teasing, and no student said that his or her friends or peers questioned their ethnic identity or suggested that they were "acting White." These responses indicate that peer support and encouragement are multidimensional and that students' peers may serve as academic resources. Even when a student belongs to a peer group whose members are not as successful as he or she is, that student's peers' response to his or her performance may often be positive. Speaking about his friends, Ian stated,

> They're not the type of people to tease anybody, so I don't think they would tease me. Maybe they would call me a nerd or whatever. 'Cause I get called that a lot by the other students, but you know, it's like I just brush it off, it don't matter to me.

It might be inferred that students are not teased because they have selected peer groups who are similar to them in terms of their attitudes and behaviors toward school; but other students pointed out that their peers were doing less well than they were for reasons that included "lack of focus" or "not doing the work."

For some students, doing well in mathematics seemed to provide some social cachet. Anita stated, "My friends like that I am a geek or genius or something." However, other students, including Adriana, noted that there is some awareness of social repercussions: "You don't want to look like a geek [by raising your hand in class to ask for help]!"

In addition to reporting positive peer responses to their academic behaviors, these high school students did report some distractions inside and outside of the classroom. Tomas said ruefully, "I try to get away from my friends in math class." What these students term "socializing" many of them viewed as potentially harmful to their current academic plans or future paths to career and college success. As Ian pointed out, "There's always peer pressure not to go to class or anything, but other than that I don't think that anyone stops anyone from doing what they gotta do."

Ian's contention that students have agency to engage in behaviors outside of ones that are championed as peer norms is in direct opposition to much of the literature, which often suggests that peer pressure is always constant and negative and, further, is something that students cannot or do not resist. Many of the students in this study were adept at resisting advances by other students to engage in classroom behavior that would interfere with their classwork.

Ellen is one student of many who recognized that some friends' academic decisions may not be aligned with her own:

> My friends are not going to be there for everything. . . . As much as I would like [to be in] a class with friends, not being in a class with them would help me focus more. It would be for my own good [to take an advanced class].

Like Ellen, other high-achieving students noted that high school was an atmosphere that facilitated socializing. Several talked about how they viewed school and the need to limit social behaviors in favor of academic ones:

> School is not a social club. This is to do what you need to do and then go. (Adriana)

> Since we are in high school, it's a pretty social environment, so there's a lot of socializing going on in class. . . . I just say OK, OK, we'll just talk after class. (Jana)

When asked if she would take an honors math class if recommended by her teacher, even if her friends weren't going to be in it, Yvette responded,

> Yes, because I'd be away from my friends. Sometimes they distract me. It would probably be more advanced, right? I'd like more advanced [work] since I'm getting it [understanding it].

Other students noted that their friends were ambivalent about teasing them about their mathematics success or talking about mathematics. Lamont reported that "most of the time they want me to help them [rather] than tease me, so, no, I get encouraged more by my friends [for doing well] than teased about it." John asserted that mathematics, while "it's not like a topic of choosing" among his friends, is important. He studies with his friend Damon. He says, "He's in my class so whatever we learn together, if we don't understand something we help each other to get it."

On the survey of Lowell High students, students of all achievement levels were asked how often they talk about grades, about school, and about going to college; as well as about studying or doing homework with their friends. They reported talking about grades, school, and going to college fairly frequently, and were less likely to study or do homework together. They reported they were slightly more likely to spend their time with friends hanging out at school or in the neighborhood than to engage in academic activities with friends.

High achievers' statements about their friends echo in general what Lowell High students think about what peers would say if they got good grades in mathematics. Among Lowell students, 56.4% say that if they got good grades in math their friends would compliment them or admire them for it, while 2.1% reported that they'd be made fun of or teased about it. Further, 17.1% of Lowell students report that their friends wouldn't care or wouldn't talk about it, while 24.3% of students aren't sure how their friends would respond to their success. While all students aren't sure that their friends would respond positively to their mathematics success, these results show a multidimensional picture of how students feel about their peers. In fact, when disaggregating the data by students' reported achievement level, there were no significant differences in how students perceived peer influences.

Ambivalence about how friends might respond to mathematics success is reflected in how high-achieving students categorize their mathematics work with friends and peers. John's contention that he and his friends don't talk about their mathematics work but yet that he and Damon work together when they don't understand something seems somewhat contradictory.[2] It may be that he does not really consider Damon to be a close friend, but rather a "mathematics class" friend. It is possible that students might have intellectual communities that help them to be successful in mathematics and that they consider to be important that extend beyond their closest friends. For example, Anita noted that "Lena [another student in the sample], she's good in math but we don't have nothing in common." Students talked about how their peers (including those who were in the same mathematics class, who did not take the same mathematics class, or who did not even attend Lowell) supported their mathematics achievement in various ways detailed below. The types of collaborative conversations that John and Damon have about mathematics are also reported by other students, in talking about their friends and peers who were taking the same mathematics course, sometimes in different class periods. Tomas reported,

> At lunchtime, I study with this girl, Katie, we do a problem set or whatever, review homework, or things like that.

Elizabeth noted,

> [My friends are] all taking the same math as me so we help each other. . . . We do homework together sometimes, and if one person, like, doesn't understand the answer, someone will, like, explain it.

When Elizabeth was asked if these friends were doing as well as she was in math and why, she responded, "Yeah, mostly. . . . We all work together." Other students described that they had conversations about mathematics homework and classwork outside of school:

> [My friends] help me with homework and stuff like that, if I need help. (Adriana)

> One of my friends called me the other night because she didn't understand her math homework. So we tried to come together and do it as best we can over the phone. (Anita)

> We'll ask each other like, "What's on the test?" or "Do you know this?" Last week there was a new girl who came in who was like, "Can you please help me with math—I really don't get it." And I'm like, I'm not really that good but I can help people. I was like, "Where are you having a problem?" And she was like, "The whole thing is just confusing to me: factoring, factoring x this and x that, difference of two squares." So we exchanged numbers and I told her that she could call me if she had a problem. (Ellen)

As these remarks indicate, many of these high-achieving students were committed to helping others and approaching each other for help if they themselves were having problems:

> Me and my friend, Andrew—he sits next to me in math, we talk about math every day after class. Like when we have a test we talk about who got the higher grade, or "why did you get that part wrong." (Linus)

> The class I'm in helps each other. (Gabriela)

> Sometimes when they [her friends in class] don't know they go and ask me or I go ask them if I don't get things. (Lourdes)

Lourdes reported that common characteristics of students who do well in mathematics at Lowell are those who "go to every class that they have. They will participate in class and help out the students who do not understand."

Students also used time outside of the classroom and school to collaborate on mathematics work. They report talking to friends who do not attend Lowell to review mathematics problems:

> A friend out of school . . . he sits with me, he tries to figure it out with me. [My classmates and I] go into the lunchroom, we talk about it. (Naomi)

> Friends who don't go to my school help me out. They're like—"Oh, I did this already at my school," so they explain to me what they have done. (Ana)

Like Davis and his friends as described in Davis et al. (2002), students report that they admonish and are admonished by other students to do their mathematics work. This exemplifies the reciprocal nature of peer support for these high-achieving mathematics students at Lowell. Said Rebeca,

> Like, my best friend, she tells me a lot, well, you need to get this together because in college you don't want to take remedial classes. You don't wanna take some class that you already know how to do it and you're going to do it again, that's terrible. You don't wanna pay for something that you don't really need. But it's like small talk, basically, we don't really talk about math.

Adriana demonstrates:

> Yeah, like, sometimes I tell my friend because I like math and she doesn't try in class and everything, and I'm, like, you need math in the future because everything's math in this world.

Another student, Craig, describes in detail how he tries to encourage one of his peers and foster better mathematics behaviors:

> Well, one of my friends, she's in my class, she gives up a lot. Like she sees a question that she can't do. . . . But she gives up real easy. If it's a question she cannot do [right away] she will look at it and she'll be like, "Oh, I can't do it," and she'll probably go to sleep, like I used to do. But I could do it, but it's just that I didn't want to. And then like, when I try to encourage her she gets mad. She's younger than me, so it seems like I'm, you know, bullying her or whatever. Like, you gotta do this. . . . I can't help her unless she sees it, or at least the way I've seen it. I tell her, "You don't wanna be like me, I'm supposed to graduate this year and I'm not going to graduate because of stuff that I used to do," and she's like, "I'm not going to be like that." OK, that's what I said. But you know, you're old enough to make your own mistakes, I guess, but I try to encourage her too. But it's a lot of people in most of the math classes that don't want to do it 'cause either one person said it was hard, or they'd rather copy from somebody else and just get it, but they'll never get it with that.

Yvette reported that she has not always been a good mathematics student, but that she wanted "to get good grades like her friends":

> It's just like seeing them going to class and seeing that, like, they prosper so much from going to class. I used to watch my friend go to class and I would do the same thing she did but I would never get the same grades. And it was

like, well, "How come I'm not getting the same grades that she does?" but um, she was actually doing all the work and I was just there, so I learned my lesson from that.

The roles of adults (parents, family members, teachers, counselors) in supporting students' networks initially seemed to be peripheral, in that the Lowell students did not often mention adults first as contributors to their mathematics success, but these roles were critical in one particular respect: All the students reported that their parents had higher expectations of their grades than did their peers. Yet this study found that parents may indeed value education but may not know how to help or guide their students in secondary school (Stanton-Salazar, 2001). This further underscores why students' connections and networks in and out of school are critically important. During her interview, one student noted that her mother's employer gave her advice about doing mathematics.

Parents' expectations of mathematics grades were very often higher than those of students, and the students' expectations of their grades were often much higher than those of their friends. Most students reported that their parents wanted them to earn grades of A or B at the lowest. No student reported that a C was an acceptable grade for themselves, although most reported that their friends thought this was an acceptable grade in mathematics. This could be due to students' differing beliefs about mathematics: It should be noted that several students talked about how "all of their families" were good in mathematics, suggesting that they might attribute their mathematics success to ability (Bempechat, 1998; Middleton & Spanias, 1999). Other students talked about the effort and hard work it took to do mathematics, thus attributing their success to effort (Bempechat, 1998; Middleton & Spanias, 1999).

In addition to collaborative activities within their peer intellectual communities, some students (predominantly male) mentioned that elements of competition help students do well in mathematics:

My friends are always competing with me, 'cause we're all smart in math or whatever, so they're always like—I'm better than you, I got higher than you, so that pushes you. . . . I like to show that I'm good in the subject. (John)

Me and my friends are constantly competing so that makes me do better. . . . I compete with them, try to always be the best one. So I do good to beat them. (Ian)

Two of my best friends who are really good in math compete all the time. (Elizabeth)

We were in the top [classes] together, so we learned together. . . . a lot of us would compete together to get the highest grade. (Linus)

These important academic conversations, activities, and behaviors in which students and their peers engaged helped these students perform well academically— whether their peers themselves were high achieving or not. Further, they served to help students maintain their high level of mathematics achievement. A number of students pointed out that their classmates exhibited a lack of understanding that math on occasion required persistence and effort, and that they wanted to serve as examples to their classmates of behaviors to avoid in order to succeed in mathematics.

How do students, in general, describe their peers' responses to academic achievement? Students were asked a series of questions about how their peers would respond to a fellow student's academic and mathematics success. Teachers were also asked these questions. Table 2.1 presents teachers' and students' responses.

The teacher sample is very small; thus comparisons between teacher and student responses should be viewed with care. However, it should be noted that a higher percentage of teachers than students believe that students will suffer a social penalty for academic success. Further, 20% of students report that their friends wouldn't care or talk about how they were doing in school, compared with only 6% of teachers. Because we were interested in evaluating if teachers and students had different standards for mathematics success versus overall academic success, we asked the same questions, but targeted them to math (Table 2.2). Responses are similar, but there are some key differences—notably that a slightly smaller percentage of students report that their friends would admire them for doing well in mathematics and a higher percentage report that they are not sure how their friends would respond. This may indicate that our societal perception of mathematics as a difficult or hard subject in which only extremely smart, "nerdy" people do well affects students' beliefs about how they will be viewed by their peers. Note, also, that 15% of students are not sure how their friends would respond to their academic success, and 24% of students are not sure about how their friends might respond to their math success. For these students, conversations about schoolwork may not occur very often.

Table 2.1. Friends' responses to students' academic success, overall, as perceived by students and teachers

Question	Percentage of students responding ($n = 154$)	Percentage of teachers responding ($n = 17$)
If a typical student at this high school got good grades OVERALL is this something that their friends would . . .		
Compliment/admire them for?	62.1	41
Make fun of them/tease them about?	2.9	18
Not care/not talk about?	20.0	6
Not sure	15.0	36

Table 2.2. Friends' responses to students' mathematics success

Question	Percentage of students responding ($n = 154$)	Percentage of teachers responding ($n = 19$)
If a typical student at this high school got good grades in MATH is this something that their friends would . . .		
Compliment/admire them for?	56.4	37
Make fun of them/tease them about?	2.1	26
Not care/not talk about?	17.1	16
Not sure	24.3	21

The students' and teachers' questionnaires also asked questions about how students spend their time with friends (Table 2.3). These questions focused on student activities outside of school and peer responses to academic engagement.

Of particular interest is the fact that Lowell teachers and students agree that students do not spend a great deal of time studying or doing homework together—despite the fact that this has been an effective route to increasing student achievement when school adults have high expectations for students that students share for themselves. It should be noted that students spend considerably more time talking about going to college than teachers think, and less time hanging out, going to parties, and working than teachers believe. It is very clear that students have a strong interest in going to college—a finding underscored by other questions and our interviews with high-achieving students.

Key Gender Differences in Descriptions of Peer Group Support

The Lowell questionnaire data do, however, support the research findings that girls and boys may have different perceptions of their peer groups. Most generally, there was one statistically significant difference between female and male students about friends' plans to go to college: Girls were more likely to report this than were boys. There were no other statistically significant gender differences on other items. But when the data were disaggregated by race/ethnicity we found that some gender differences are masked. For example, when data for Black and Latino/a students are examined for gender differences, differences in attitudes toward school, mathematics, academic behaviors (as we saw in Chapter 1), and peer influences emerge. For example, African American boys report significantly more negative peer responses to high mathematics achievement than do African American girls. They report hanging out more in the neighborhood with friends than do African American girls. Latino boys also report more negative peer responses than do Latina girls, but for academics overall. Latino boys are also significantly less likely than Latina girls to report studying together with friends.

Table 2.3. How students spend time with their friends

Question	Mean student response ($n = 154$)	Mean teacher response ($n = 23$)
With closest friends, how often do students spend time . . . ? (1 = never, 5 = a lot)		
Talking with each other about grades	3.2	3.4
Studying/doing homework together	2.4	2.6
Talking about going to college	3.5	2.6
Hanging out in the neighborhood	3.6	4.4
Going to parties, clubs, concerts, etc.	3.3	4.2
Working to get extra money	3.1	3.9

When we examine racial/ethnic differences among boys and girls, we also see some interesting patterns. For boys, Black students report more often hanging out with their friends (whether it's in the neighborhood or to study or do homework together) than do Latino students. Black males also are more likely to agree with the statement "Most of my friends like math" than are Latino males. For girls, Black students report that they "talk about school" with their close friends more often than do Latina students.

CONCLUSIONS

These urban high school students, like those described in the research literature, report having peers, classmates, close friends, parents, older adult relatives, siblings, or cousins who contribute to their mathematics success. It is very important to note that high-achieving students in particular are influenced by family members other than their parents. Their "near peers" within the family—cousins and siblings were frequently mentioned—are important contributors to students' mathematics success. In turn, the Lowell students influence other family members, often younger siblings, cousins, and fellow classmates. Further, many of these students describe their interactions with others in ways that suggest that they model good academic behavior for other students who are not related to them, learning from their own out-of-school role models. These findings suggest that these students may have benefited from the behaviors of their own academic communities—which include parents and other family members and near peers outside of school—and disperse their behaviors through relationships, conversations, and behaviors with their own intellectual communities, comprising fellow students, siblings, and cousins. Thus, there may be ways that positive academic behaviors can be dispersed throughout schools through the fluid relationships that students have with close friends, peers, and classmates, which could be integral in improving the overall academic climate and achievement of schools. In addition, our understanding of students' peer communities has implications for how we view students' peer influences: The extant research and the Lowell data make clear that students make important choices about how they navigate expectations from peers and that these expectations are multidimensional and complex. Students may choose to downplay their academic behaviors with some peers and not others. Both low-achieving and high-achieving peers can have positive effects on students' performance and learning behaviors.

It is important to note that numerous college interventions and initiatives—both formal and informal—targeting underrepresented students in mathematics use academically supportive peer groups and their academic activities outside of the mathematics classroom to promote access to the sciences and higher mathematics achievement for underserved students. What we know about high-achieving students and the institutions that facilitate achievement can and should inform interventions designed to improve student outcomes in mathematics.

Several college initiatives, notably Uri Treisman's Mathematics Workshop Program at the University of California at Berkeley (Fullilove & Treisman, 1990; Treisman, 1992); Clarence Stephens's programs at Morgan State College (now University), State University of New York (SUNY) at Potsdam, and SUNY Geneseo (Datta, 1993; Megginson, 2003; Walker, 2010); and the Meyerhoff Scholars Program at the University of Maryland in Baltimore County (Hrabowski, Maton, & Greif, 1998), have helped to support and develop mathematically successful students from underrepresented populations. A critical component of these three initiatives (which do not constitute an exhaustive list of successful programs for underrepresented students in the sciences in college) is the use of academically oriented peer groups.

The creators of these programs sought to address mathematics learning outcomes among underrepresented students of color who demonstrated high achievement. At Berkeley, Treisman in particular sought to learn not only why Black and Latino/a students were failing calculus but also why the Chinese American calculus students were succeeding at higher rates than those of others. He discovered, as have other researchers, that Asian American students spend substantially more time and effort on homework and other academic activities than do students from other ethnic groups; their parents emphasize effort and hard work as being necessary to excel academically, rather than encouraging children to rely on one's "natural" ability; and students frequently work in groups on academic tasks outside of school. These reasons for Asian American success in mathematics are clearly not genetic, but are rooted in environmental and cultural practices that support talent development.

Clarence F. Stephens was the ninth African American to receive a PhD in mathematics (University of Michigan, 1943), was a professor of mathematics at Morgan State from 1947 to 1962, and later had an equally influential career at

SUNY Geneseo and Potsdam. Throughout Dr. Stephens's career, he has been known for increasing the number of students interested in mathematics and encouraging students to pursue advanced degrees in mathematics and related fields. One of Dr. Stephens's former students (now a mathematician himself) describes the program at Morgan:

> Stephens's program was really, I think, rather innovative. First of all, our students worked with each other a lot. We even tutored students who weren't as good as [we were]. We worked with each other a lot, and then we found ourselves competing for things. But even from our sophomore year, he'd bring in a copy of the *American Mathematical Monthly*. *Mathematical Monthly* had two types of problems: advanced problems and elementary problems. Supposedly, elementary problems were solvable by senior math majors and early grad students. But the advanced problems were more for serious mathematicians, professional mathematicians. But he would bring these problems in and we would just work, trying to solve all of them. (Dr. Scott Williams, personal communication, 2008)

One important feature of all three programs is the extent to which student participants worked together and supported one another academically. They also demonstrate that it is very important for students of color to experience mathematics success in the context of working together as a group, rather than working in isolation. This is much less of an issue for White and Asian students, who make up the vast majority of students who enroll in high-level mathematics classes from middle school through college and beyond. Thus, underrepresented high-achieving students of color experience a degree of isolation that may make it difficult for them to persist, or want to persist, in high-achieving behaviors (Fries-Britt, 1998). Having "academic relationships" with one's peers is a critical part of student academic talent development.

Another critical feature of all three programs is that in addition to building on and facilitating students' peer networks, they begin from the premise that underrepresented students can be excellent mathematics students. This feature is something largely missing from the discourse about urban high school students. In the next chapter, I discuss the important role of schools and teachers' discourse and actions in fostering mathematics success for high school students.

Facilitating and Thwarting Mathematics Success for Urban Students

Despite decades of reform, efforts to improve national mathematics achievement scores, particularly for urban students, have largely been unsuccessful. While the underperformance of American students in general in mathematics is a national problem, it is clear that students attending urban schools are the most under-performing (Thirunarayanan, 2004). Why? In the Introduction and Chapter 2, we discussed persistent misperceptions of urban students and mathematics; the complexity of the "achievement gap"; and how attitudes about urban students, mathematics, and performance gaps might have an effect on school practices and policies that affect mathematics opportunities. In Chapter 2 we learned that despite many urban students' positive attitudes and evidence of supportive com-munities for mathematics and education, these positive beliefs don't "add up" (Ladson-Billings, 1997) to better performance in mathematics. While it is cer-tainly true that students' own behaviors and beliefs matter, these are affected by and affect teachers' and administrators' behaviors and beliefs within schools. In addition, policies external to schools have significant impact on issues of teach-ing and learning in classrooms. In this chapter, I discuss more specifically the many ways that students' mathematics success is both thwarted and facilitated by national, state, and school district policies, as well as by policies and practices within schools, departments, and classrooms.

In my years as a mathematics student, teacher, and researcher in urban settings, I have collected "cases" of mathematics policies and practices that thwart math success for urban students. Here are two examples drawn from districts, schools, and classrooms:

> In one urban school district, students are tested to determine entrance into honors mathematics courses. Although there is a substantial Spanish-speaking population, students who enter the district and speak only Spanish are expected to enroll in a "transition" general-level mathematics class. Further, they are not allowed to take the entrance exam in Spanish.

A Black student excelled in her general-level algebra course. At the end of each quarter students are evaluated, on the basis of grades, to determine if they should be moved into a higher- or lower-level course. When an administrator asked why the student had not been moved to a higher-level course her teacher replied that she needed the student to remain in the course because she was a good influence on, and a good role model for, the other students in the class (who were predominantly Black and Latino/a).

These examples show that district-, school-, and classroom-level decisions all can have a major impact on students' access to mathematics. Troublingly, these examples are rooted in limiting beliefs about the potential for and necessity of high mathematical performance of students of color in mathematics.

The testing example is common practice in many school districts. The implicit assumption that all students who enter school from Spanish-speaking countries need remediation in mathematics does not allow for the possibility that one could be Spanish speaking and mathematically gifted. Not allowing students to test in their language to determine course placement, particularly for mathematics, seems to ensure that these students are consigned to show poor performance in mathematics. For placement purposes, it is perfectly logical to test entering students in their native languages to determine their mathematical abilities, without the confounding element of testing their English proficiency also. Robert Crosnoe (2006) points out,

> Even more worrisome is the potential for US educators, on the basis of their early evaluations of these entry level skills, to shape the instruction and placement of children in self-fulfilling ways. For example, a teacher views a gifted Mexican immigrant child to be unintelligent because of her difficulty speaking English, and consequently recommends that this child be placed in remedial coursework that provides no intellectual stimulation or challenge for that child and eventually causes her to disengage from school and do poorly. In this way, the low level of English proficiency and early math skills characteristic of children from Mexican immigrant families could even trump their actual aptitudes and abilities. (Crosnoe, 2006, pp. 38–39)

The classroom example assumes that this particular Black mathematics student's sole importance is to help the teacher maintain order in the classroom. In essence, the teacher deemed that it was more important for the student's meritorious achievement in mathematics to help the teacher than to advance the student. Such practice holds the student back—contributing, despite the student's excellence, to continued underrepresentation of Black students in high-level mathematics classes.

Most disturbing is that these examples, as do many school, administrator, and teacher practices, can have cumulative and long-lasting effects on students' mathematics and academic outcomes. Although there are many students of color who have persisted and excelled in mathematics despite such experiences, these kinds of obstacles must be removed.

But it is important to note that what happens at the school and mathematics classroom level operates within a larger policy realm. Decisions made and policies implemented at the national, state, city, and school district levels contribute to decisions made about mathematics in schools and classrooms and can either facilitate or impede student performance in mathematics, particularly for urban students.

NATIONAL, STATE, AND SCHOOL DISTRICT POLICIES AND PRACTICES

In the United States, unlike many other countries, education is heavily under local control, meaning that states and local school districts make decisions about schools and students. Although the federal government has been involved in education in some way in various incarnations since the formation of the Department of Education in 1867 (to collect information on schools and teaching), the creation of the Department of Health, Education, and Welfare in 1953, the subsequent formation of the Department of Education in 1979, and the authorization and reauthorization of legislative acts designed to fund compensatory and early education initiatives like Head Start in 1965 and the Elementary and Secondary Education Act of 1965 (Lappan & Wanko, 2003), generally the federal government has not ventured into local policy issues. However, as Schoenfeld (2002) notes, in times of national crisis increased attention was paid to education, particularly mathematics and science education. For example, the launch of *Sputnik* in 1957 resulted in concerns that the United States was falling behind other countries—specifically the Soviet Union—scientifically, and this generated significant funding for and curriculum development in mathematics education. The National Science Foundation became increasingly involved in funding mathematics and science education programs, initiatives, and curricula. National proclamations and policy statements around education became increasingly frequent in the 1980s and 1990s, especially those targeting mathematics and science and improved performance of U.S. students relative to international counterparts (Schoenfeld, 2002; Lappan & Wanko, 2003).

However, many suggest that with *A Nation at Risk* (National Commission on Excellence in Education, 1983), a report on the state of U.S. education commissioned by then secretary of education Terrel Bell in 1983, the U.S. government began to have increased influence on local educational decisions that eventually led to the passage of the No Child Left Behind Act of 2001 (NCLB), a legislative act whose stated purpose was to "close the achievement gap with accountability, flexibility, and choice, so that no child is left behind." NCLB has had immediate and far-ranging, national and local, intended and unintended consequences on policies and practices around curriculum, instruction, and assessment.

Related to mathematics, there were a number of policies emerging from the 1980s with the advent of the *A Nation at Risk* report. Namely, *A Nation at Risk* made numerous policy recommendations, including one focused on increasing high school graduation requirements. For mathematics, *A Nation at Risk* suggested that at least 3 years of high school mathematics was necessary to adequately prepare students for college or postsecondary work. With this increase in mathematics requirements, questions about whether the mathematics being taught in secondary school was adequately preparing high school students soon emerged. National initiatives around "Algebra for All" developed, and a number of states began to mandate that at least one of the courses taken during the recommended 3 years of high school mathematics be algebra, which led to a reduction in the availability of general mathematics coursework for most students. The National Research Council released *Everybody Counts* in 1989, which argued for mathematics education reform from kindergarten through graduate school. The National Science Foundation–funded Urban Systemic Initiatives (later Urban Systemic Programs) began in 1994 and targeted mathematics and science education in urban school districts, eventually reaching nearly 13 million students (Stern & McCrocklin, 2006). Other entities entered the mathematics education policy arena; one example of a national initiative was the College Board's Equity 2000 program, which was developed to address the racial/ethnic participation gap in algebra and advanced mathematics course-taking and the subsequent gap in college-going rates by mandating that participating school districts enroll all students in algebra by the end of grade 9, and geometry by the end of grade 10 (Ham & Walker, 1999).

But many observers note that the increased attention paid to standards in education proposed by *A Nation at Risk* took a particular turn when the National Council of Teachers of Mathematics (NCTM) presented its *Curriculum and Evaluation Standards* in 1989. As Schoenfeld (2002) notes, the publication of the *Standards* and subsequent documents targeting assessment and teaching "catalyzed a national standards movement" (p. 15). Other disciplines, science and English, for example, followed the NCTM in drafting standards documents. The NCTM's *Standards* and subsequent *Principles and Standards* (2000) have had a dramatic impact on curriculum, textbooks, assessments, and teacher education. The "turn" in mathematics education from a teacher-centered, didactic, skills-based paradigm to a student-centered, process-oriented, concept-focused one is due in large part to the NCTM *Standards*. But differences in implementation of the suggested mathematics education reforms have been and are profound. As Tate (1994) noted:

> The call for new mathematics standards represents an epistemological shift in school mathematics from a shopkeeper (basic skills) philosophy of mathematics pedagogy to a constructivist, technology-driven vision of mathematics instruction (Romberg, 1992). The fiscal appeal of a basic skills mathematics curriculum is low implementation cost. However, the standards-based vision requires urban and rural schools to (a)

reallocate funds, seek additional funding to improve teachers' mathematics knowledge, or both; (b) update instructional materials (e.g., textbooks, laboratories, and computer facilities); and (c) enhance the quality of other resources (e.g., deteriorating classrooms and buildings) (Darling-Hammond, 1995; Kozol, 1991). (p. 675)

Tate points out that resource differences matter and affect both the implementation of reforms as well as opportunities for students to benefit from the reforms. Although the framers of mathematics education reform initiatives in the past 3 decades have suggested the need for increased equity and access in mathematics education for underserved students, policies that address the stark differences in educational opportunity between more affluent and less affluent students have not been implemented widely.

Proclamations, standards documents, and policy statements since *A Nation at Risk* have addressed the issue of equitable opportunities and outcomes for students; in particular, the NCTM's *Curriculum and Evaluation Standards* (1989) as well as the *Principles and Standards* (2000) have addressed it prominently. These documents have highlighted the importance of equity in addressing and ameliorating disparities in opportunities and outcomes for underserved students. Many of these documents describe forthrightly how mathematics has traditionally been the purview of the elite, and specifically, the domain of White males. Indeed, there was not a concern, historically, that women or members of certain ethnic minority groups were not represented in mathematics fields. The shifting global terrain, however, has underscored the need for American talent to be developed, wherever it can be found. Most of the standards documents make specific links to national prominence in international arenas and improved outcomes for all U.S. citizenry.

NCLB's ongoing effects will continue to be debated for a long time. Most observers acknowledge that it was long overdue that something was said and done about persistent race/ethnicity gaps in student performance. NCLB tacitly acknowledged that some of these gaps were due to factors controlled by schools, suggesting, for example, that every student should be taught by a highly qualified teacher, defined by NCLB as an individual with a bachelor's degree, full state certification or licensure, and proof of subject knowledge. However, interpretations of this last are varied, and there was little done to address the need to recruit and retain effective and well-prepared teachers. Further, NCLB's focus on reading and mathematics has often meant that other subjects—including science, social studies, art, and physical education—get short shrift in schools. In addition, NCLB's focus on schools' demonstrating annual yearly progress to remain in good standing meant that states were testing students more often, and this in many cases led to a return to testing that focused on basic skills and reduced measures that were designed to assess students' conceptual understanding and problem-solving ability, something which had been urged by the NCTM and other groups.

Finally, many have suggested that national reforms have limited impact and are largely unsuccessful in meeting their goals, because they do not attend sufficiently to issues of teaching and learning at the district, school, and classroom level (Elmore, Peterson, & McCarthy, 1996). A common critique of the initial NCTM *Curriculum and Standards* was that they were "long on direction, and short on detail" (Schoenfeld, 2002, p. 15).

With the advent of the NCTM *Standards,* states began to develop their own curriculum standards and frameworks for mathematics, aligned with the *Standards*. Accordingly, states also began to develop their own standardized assessments to measure students' achievement and progress (although some states had already been doing this—New York State's Regents examinations, which date to 1866, are an example, although the modern incarnations of these exams date to the 1970s). Some states' assessments were focused on basic skills, while others designed assessments that incorporated measures of skills as well as problem solving. With wide variety across states, the National Governors' Association Center for Best Practice and the Council of Chief State School Officers began work on the Common Core State Standards in the early 2000s. In part a response to the proliferation of multiple sets of states' standards around disciplines, it remains to be seen what impact the Common Core Standards will have on mathematics curriculum, instruction, and assessment in schools. But if *A Nation at Risk,* the No Child Left Behind Act, and the NCTM standards documents are any indication, we can expect impact in ways that we can foresee and also in ways that we cannot. It is also important to note, as Tate (1997) and others remind us, that the presence of standards is perhaps a necessary but not sufficient condition for quality, equitable mathematics teaching and learning in schools.

SCHOOL RESOURCES AND ORGANIZATIONAL PRACTICES

Reformers have noted that, despite federal, state, and district policies designed to improve academic and mathematics outcomes, major educational inequities within schools continue to have a significant impact on mathematics outcomes. In addition, the promise of the U.S. Supreme Court's *Brown v Board of Education* decision in 1954 to alleviate school segregation—broadly and narrowly construed—and its attendant state-supported resource disparities, particularly in the South, has largely been an elusive one. The Civil Rights Project, initially at Harvard and now housed at the University of California, Los Angeles, has found in the past 2 decades an increase in school segregation, to the same levels or worse than before the *Brown* decision. In deeply segregated schools, students of color and low-income students may be less likely to be exposed to highly qualified teachers, extensive resources, or a network of challenging mathematics courses than are their White and more affluent counterparts. In integrated schools, African American and Latino/a students may find themselves, whether their achievement levels are high or low, tracked into lower-level mathematics courses than are their Asian or White counterparts. These experiences reflect lowered expectations and translate into lowered outcomes, ironically approximating the experiences of their counterparts who attend segregated schools that do not offer challenging coursework. In particular, Loveless (1999), Ogbu and Davis (2003), and others have noted that "teachers of students in lower track courses assign less homework and focus their instruction more on basic comprehension and memorization of facts than on critical thinking" (Harris, 2010, p. 1170). Thus, regardless of what kind of school Blacks and Latino/a students attend—whether racially desegregated or segregated, or urban or suburban—they may be receiving an education substantially different from that of their White peers (Darling-Hammond, 2004).

Tracking is probably the most well-documented component of this dual system of education, and is the major cause of racial inequality in educational opportunity and consequent opportunity in later life (Haney, 1993; Oakes, 1985, 1995; Ogbu, 1978). Despite court cases in recent years challenging the use of tracking in schools because it has a disproportionate, deleterious effect on students of color, Black and Latino/a students are "still being grouped and treated in ways that constrain their educational opportunities, despite their attending nominally desegregated schools" (Secada, 1992, p. 625).

Indeed, Ogbu (1987) notes that although minority and White children in desegregated schools have the same teachers, administrators, and services, a subtle mechanism that maintains Black students' position in lower-level classes is the "lowered expectation of teachers and administrators" (p. 90). Because "schools far more often judge Black and Latino/a students to have learning deficits and limited potential" (Oakes, 1995, p. 682), they are more likely than Whites to be placed in lower-level academic classes. Oakes's research concerning the educational opportunity of Black and Latino/a students within racially diverse schools documents that Black students are disproportionately overrepresented in lower academic tracks and disproportionately underrepresented in higher academic tracks (Oakes, 1995; Donelan, Neal, & Jones, 1994; Secada, 1992; Oakes, 1985). Although the overrepresentation of Black and Latino/a students in lower-level courses may be due in part to reliance on standardized test scores to determine track placement, Oakes (1995) and Useem (1990) found that often subjective, noneducational criteria such as teachers' evaluation of student behavior and parental influence rather than performance on standardized tests determine track placement. For example, Oakes (1995) found that despite the emphasis on using standardized test scores in one district to assign students to high- and low-level mathematics and English classes, "African-American and Latino students were much less likely than White or Asian students with the same test scores to be placed in accelerated courses" (p. 686).

The difference in education that students in low- and high-track classes receive is clear. Students placed in "lower-ability," or non–college preparatory, tracks take courses that are less academically demanding and take them with teachers who expect less learning from their students (Haney, 1993; Oakes, 1985). Low-track mathematics courses, for example, are often assigned to beginning teachers or to "teachers who have fallen out of favor with a school administrator" (Secada, 1993, p. 646). Conversely, students in high-track courses "often benefit from enthusiastic and highly motivated 'master teachers' and stimulating environments that students in the lower tracks seldom see" (Donelan et al., 1994, p. 383). Indeed, teachers "do establish differential patterns of interaction with their students based on student demographic characteristics as well as on their expectations of student success" (Secada, 1992, p. 644).

Finally, once students find themselves on a particular track, mobility from a low to a high track is virtually impossible. The longer students stay in school on a low track, the wider the gap between their learning and that of students in high-track courses becomes. Thus, the discriminatory placement of students in low tracks can exacerbate racial differences in achievement scores because it creates a cycle of restricted opportunities and diminished outcomes (Oakes, 1985). There is evidence that for some schools instituting detracking policies in mathematics results in improved gains for minority and low-SES students and a narrowing of ethnic and socioeconomic gaps in participation and performance (Burris, Heubert, & Levin, 2006). In addition, these studies addressed a common concern of tracking proponents: that high-achieving students' performance was not harmed by the introduction of detracking policies.

But in addition to differences in opportunities within schools, there are disparities between schools as well. Much of the research about mathematics teaching in urban schools suggests, unfortunately, that it follows a "pedagogy of poverty" (Haberman, 1991). Teachers' and other school adults' perceptions of urban, poor, and minority students can contribute to how teachers structure learning opportunities in school (Bell, 2002; Jamar & Pitts, 2005). As Ladson-Billings (1997) and others have found, teachers who have low expectations of urban and minority children adopt a "pedagogy of poverty," in which basic skills instruction is the norm. Gilbert (1997) found that prospective teachers felt that "a basic skills focus was forced by the way [urban] students behaved, by their 'attitude,' and by the kind of content urban students needed" (p. 90). While "Anyon (1997) and others have warned of the dangers of limiting the education of children of poor and working-class families to highly structured rote teaching and learning" (Breitborde, 2002, p. 42), in urban schools teachers often use "mathematics instruction centered on basic skills and repetition, rather than instruction that provides [students] with opportunities to learn and exercise higher order thinking skills" (Walker, 2003, p. 18). Providing rigorous, high-quality mathematics instruction for all students has become critically important, to ensure that students will be able to participate in economic, career, and educational opportunities for which strong mathematical preparation is a prerequisite (Moses & Cobb, 2001). In mathematics, especially, teachers in urban schools who have low expectations and stereotypes about urban students may feel that this basic and rote instruction is best, despite the advances in mathematics curricula and pedagogy as espoused by the NCTM (Jamar & Pitts, 2005; Ladson-Billings, 1997). Further, the need for high-quality mathematics instruction is critical for students to be prepared for full access to educational, career, and economic opportunities.

Despite parents' interest in and commitment to their children's education, as described in Chapter 2, there is evidence that urban schools are not particularly welcoming or receptive to students' parents. Substantial research details the importance of parental involvement in student achievement and evidence that parents of students of color are committed to the academic success of their children. However, schools are often unwelcoming or hostile to parents and do not wish to include them in the planning of school events and academic decisions about their children (Auerbach, 2002; Delpit, 1995). School administrators, teachers, and counselors may consider parents of students of color to be obstacles to overcome, citing that the parents are uneducated and not interested in students' academic work (Yan, 1999). For example, school officials point to low attendance on parent-teacher nights in urban schools as an indicator of parents' disinterest in their children's education, without recognition of parents' work and child care schedules and without acknowledging efforts by parents to stay connected to their child's school (Cooper, 2003). Too often, school practices and culture are not conducive to enlisting the aid of students' parents to support academic engagement (Auerbach, 2002; Delpit, 1995). Well-connected, affluent parents are often more valued by schools (Valenzuela, 1999), and parents whose socioeconomic and ethnic background differs from that of administrators often find themselves outside of important decision-making processes about their own children (Auerbach, 2002). Research shows that parents of color and those who are of low SES may be less likely to question school decisions about their children, stating that "the teacher knows best" (Polite, 1994; Useem, 1992).

But even when parents are active and engaged, and are encouraged to be actively involved in and are welcomed by schools, it's clear that many decisions about mathematics opportunities have already been made at a higher level. One example of how state and school district policies can affect practices in school is found in the study completed by Useem (1992) on the offering of calculus in Massachusetts. School administrators in 26 school districts were interviewed about policies related to ability group and track placement in mathematics. Two of the school districts, geographically adjacent and similar in demographic characteristics, were studied further to understand how school district administrators made policy relating to advanced mathematics course-taking. As Useem (1992) notes, there were diverse attitudes and practices about the importance of calculus in the high school curriculum, as well as student course placement policies, sometimes within the same districts. Here are some of the example of some of the idiosyncratic policies on grouping and offering calculus to high school students:

It is better to make a mistake and err in the direction of holding a kid back. . . .
We *like* to overpopulate the seventh grade pre-algebra so the students will have an
opportunity. . . .
Some people say it is a working class community and we shouldn't have too many
people in accelerated math. . . .
Calculus is only for a few kids. . . .
Students are much better off with calculus in high school . . . .
I favor having kids get into calculus. We should find ways to help them get around
roadblocks. . . .
Teachers don't always agree but they know my philosophy and do it. . . .
I'm more important than the principal. (quoted in Useem, pp. 340–341)

Whether or not calculus is offered by schools is critically important, as Escalante
and Dirmann (1990), Useem (1992), and Werkema and Case (2005) describe. It sends
a positive signal to college admissions officers and "it sends an important message
[to students, teachers, parents, and others] about the academic tone of a high school"
(Werkema and Case, 2005, p. 512). In Werkema and Case's description of a "turnaround" urban high school, they focus on the administrators' decision to offer AP
calculus. The authors note that the administrators felt that "the fact that the school had
posted low achievement test scores for less advanced levels of mathematics was no reason to postpone the introduction of calculus or to think that no students at the school
were ready to take on that challenge" (p. 513). In the words of the assistant principal:

See, I'm not going to wait for the SAT scores to go up. . . . I think we have to
look at the best interests of the children where they are. . . . I'm not saying all
of the kids are going to end up in AP Calculus. But at least if they would like
to pursue more math, go on to higher courses in math, they're here.

These examples show that parents', students', and teachers' behaviors and decisions
about mathematics operate in a broader and significant context.

The importance of teachers and school counselors in ensuring student access
to rigorous and meaningful mathematics opportunities cannot be underscored
enough. Several studies of college students, college graduates, graduate students,
and science, technology, engineering, and mathematics (STEM) professionals discuss the role of teachers (both STEM and non-STEM), counselors, and administrators in fostering school success. In a study of African American males pursuing
engineering majors in college, Moore (2006) found that teachers and counselors
had critical conversations and provided opportunities for students to develop their
mathematics interests. In short, as described by Moore (2006); Hrabowski, Maton,
and Grief (1998); and others, the work of school adults extends beyond the classroom. These individuals may encourage students to complete particular courses,
but they are also gatekeepers to enrichment activities that facilitate mathematics
interest and engagement, as will be discussed further in Chapter 4.

DEPARTMENT POLICIES, STUDENT LEARNING, AND ACHIEVEMENT

Effective departmental structure and practice have been found to have a positive link to student achievement (Lomos, Hofman, & Bosker, 2011). A number of mathematics education researchers have examined mathematics departments, specifically, sometimes within a lens of mathematics reform practices.

In a study of "mostly urban" mathematics departments effective in advancing Black and Latino/a students, Gutiérrez (2002) found that student success is facilitated in mathematics departments that are "organized for [student] advancement," meaning that those departments have "practices, beliefs, and a general culture that maximizes the number of students choosing to enroll and successfully complete mathematics courses, especially those at the advanced levels" (p. 68). The five key characteristics of such departments are a rigorous curriculum, active commitment to students, commitment to a collective enterprise, a resourceful and empowering chairperson, and standards-based instructional practices. Particularly relevant to Latino/a students, she found that a high school department comprising two Anglo teachers and one Latino teacher was effective at advancing a significant number of the school's Latino population to Advanced Placement calculus. Through the other characteristics of the departments detailed earlier, but also through several practices that related to students' languages and culture, the department enacted a philosophy in which they built on students' strengths, rather than focused on perceived deficits of their students.

Like the Burke School in Werkema and Case's (2005) study, the department in Gutiérrez's study saw the presence of AP calculus as a critical milepost for students and teachers. Thus, the department faculty required all students to take algebra in their freshman year and geometry in their sophomore year. Numerous summer opportunities—for acceleration rather than remediation—to take college-level courses in algebra were offered. Students who were destined for calculus in their senior year took a double period of calculus (precalculus and AP calculus) to prepare for the AP exam. Students who did not take calculus were still encouraged to take college algebra or statistics or some other math class in their senior year.

What may be of most importance in Gutiérrez's work is that despite the common philosophy of the department's teachers, they enacted diverse, but effective, classroom practices. These practices centered around "encouraging students to work in their primary language and building on students' previous knowledge and stressing the language of mathematics" (Gutiérrez, 1999, pp. 1076–1077). Teachers enacted classroom behaviors in multiple ways that supported these themes.

Further, Boaler (2006) and Boaler and Staples (2008) found that reform-oriented practices focused on equity, as endorsed by a mathematics department in an urban high school, contributed to student success on problem-solving assessments as well as their enjoyment of mathematics. As Boaler (2006) describes, the department employed a variety of strategies to foster student success, including heterogenous classes instead of tracked courses (using an approach called "complex instruction" designed to counter social and academic status differences), group-worthy problems (as part of a curriculum designed by the department, in which students almost always worked in groups), shared responsibility (in which students were responsible for the learning of their peers), block scheduling (generally, 90-minute classes were taught in half a school year, rather than over a full academic year), and departmental collaboration (many hours were spent by mathematics teachers planning curriculum, designing problems, and sharing pedagogical strategies). While teachers prized the principle of equity very highly, "they did not use curriculum materials that were designed specifically to address issues of gender, culture, or class" (Boaler, 2006, p. 368).

CLASSROOM DYNAMICS: TEACHERS, STUDENTS, AND THEIR INTERACTIONS

Research around reform efforts to improve equity and increase student achievement also targets the necessity of changing teacher behaviors (largely through curriculum and assessment) in mathematics classrooms (Walker, 2009a). Yet these largely one-dimensional efforts do not critically or emphatically address a crucial dimension of the mathematics classroom: the teacher-student *interaction* in terms of valued norms, behaviors, and instructional practice. As discussed in Chapters 1 and 2, teacher beliefs about mathematics and who can do it, and teacher expectations of certain students' mathematics ability, contribute greatly to the opportunities that teachers provide students in their classroom and those students' responses (as measured by academic behaviors, for example) to the presence/absence of opportunities to learn. Certainly, the curricular and organizational mandates designed to enhance instruction are important to examine through the lens of teaching and learning. Teachers largely enact the curriculum, and students respond to how it is enacted.

It is important to note that the courses we mandate are not taught by automatons, nor does school curricular and organizational policy happen in a vacuum. The courses that are offered, and the level at which they are offered, reflect school adults' thinking about the students they teach and of what they think those students are capable. The students sitting in these classrooms are not automatons, either. Their academic and nonacademic behaviors—or, as important, their *perceived* behaviors—affect how teachers structure and deliver curriculum. In short, student access to quality mathematics depends on what the school adults in a system or school think about their students' capacity for learning mathematics. There is substantial evidence that we don't think very highly of the capacity of urban and low-income, ethnic minority students' ability to do well in mathematics unless they are Asian or Asian American. When teachers say, "Oh, I could do that problem with my advanced kids but not my low kids," or "Just do the first 10 problems—they're the easiest," or "We're not going to cover proofs in this geometry class," they are making critical decisions about the mathematics content that their students will receive.[1] In their case study of elementary mathematics teaching, Cahnmann and Remillard (2002) found that when confronted with students' confusion about a particular task, a teacher "tended to reduce the complexity of the task" (p. 199). When teachers engage in these practices, they are making critical decisions about their students! These practices, based on their expectations, have lingering and critical effects on the school and life outcomes of students.

If the choices that teachers make about instruction were quantified in some way, undoubtedly these decisions could explain some of the gap in mathematics performance we see between Black and Latino/a students on one hand and Asian and White students on the other. On many occasions too numerous to count I have heard teachers and administrators say that "these kids" just aren't motivated, that they have difficult home lives, that their parents don't care, that they don't have anyone to help them with their homework. Regardless of whether or not these statements are true, it is curious that the response is often to offer less in terms of mathematics instruction instead of more.[2]

When we consider instruction, then, it is imperative that we note that teachers' beliefs, knowledge, and attitudes about the subject matter and how to teach it are filtered through their beliefs about students and their potential. Providing equitable instruction for students will require that all of us—researchers, educators, policymakers, teachers, parents, administrators—consider, examine, and address the embedded relationships between what is done in the classroom and our expectations of students, their performance, and their possibilities. Without this work, we will continue to enact at best piecemeal and at worst woefully incomplete solutions to a complex problem, and equity will continue to elude underserved students.

If students or teachers (or other school adults) have low expectations for academic performance from Black and Latino/a students, this can be exhibited in many ways. The dynamic teacher–student relationship posited by Reyes and Stanic (1988) and elaborated upon by others suggests that students may respond to low expectations (manifested in less rigorous coursework and lack of classroom management) by not focusing on academic work or by being disruptive in class. However, teachers may respond to perceived lack of interest on the part of students by not presenting challenging, interesting, or motivational material. This cycle easily results in lowered expectations for achievement and progress—thereby affecting student performance on examinations and persistence in math classes.

One, then, could expect from the literature that students' attitudes toward school and mathematics, and perceived and actual teacher beliefs, all interact to influence student and teacher behavior. These dynamic relationships are influenced by peer behaviors and norms, which may be mediated by race/ethnicity and gender. For example, because students' peer group membership is often related to their gender and ethnic identity, that group's norms and behaviors, rooted in cultural and social contexts, can have a strong impact on students' academic attitudes and expectations and behaviors. When teachers engage with students, their academic attitudes, expectations, and behaviors may be affected by students' gender and ethnicity and teachers' perceptions of people who share those demographic characteristics. Further, the relationship between teachers' attitudes, expectations, and behaviors and those of students is a reciprocal one: Teachers and students respond accordingly to how they perceive the other thinks about them. For example, "Flowers et al. (2003) found that students' perceptions of how teachers perceived them had a profound impact on their educational aspirations" (Morris, 2006, p. 259).

Substantial research reveals that school student body characteristics are strongly correlated with opportunities to enroll in high-level, rigorous, college-preparatory courses (e.g., Oakes, 1990; Darling-Hammond, 1995; Gandara, 2004). But why? When school adults work at an urban, predominantly Latino/a or African American school, for example, it is possible that their attitudes and beliefs about Latino/as and African Americans affect how they interact with students on an academic level as well as a social one. Indeed, school adults' beliefs about the prognosis for performance for students significantly affect policy decisions around curriculum, school organization, extracurricular activities, and discipline (Allexsaht-Snider & Hart, 2001; Gutiérrez, 2000; Yonezawa et al., 2002). In these settings, academically supportive peer groups for students of color might exist but could be faced with school structures and policies that make it difficult for students to maintain connections to academics and may, indeed, help to perpetuate peer groups' negative social consequences for student achievement.

At Lowell, despite teachers' commitment to students, budget and size constraints have meant that opportunities to take advanced mathematics classes are often limited. During the 2004 interview study at Lowell, two students mentioned that there were not advanced mathematics classes at Lowell beyond the Regents-prescribed Math A and Math B courses. Said Tomas, "We [he and his friends] always talk about college and math. We're trying to get a pre-calc or calculus [class] for next year here [at Lowell]." Unfortunately, during the 2010–2011 school year, there were still limited advanced math course offerings.

The questionnaires with Lowell students revealed that they thought highly of their teachers, and interviews with students revealed that they thought deeply and with complexity about the work of teaching and what they considered to be effective teaching. Central to students' experiences with mathematics in urban high school classrooms are their teachers, their teachers' beliefs about them as students and their potential, their teachers' beliefs and attitudes about mathematics, and their teachers' instructional practices. Even if students have supportive family and peer communities, their teachers—and the work of their teachers—are critical to their mathematics success.

EFFECTIVE MATHEMATICS TEACHING

The work of teaching is complex. As Hiebert and Grouws (2007) note, "Documenting particular features of teaching that are consistently effective for students' learning has proven to be one of the great research challenges in education" (p. 371). In their examination of the research literature on mathematics teaching, Hiebert and Grouws suggest that "different teaching methods might be effective" (p. 374) for different learning goals and that classroom dynamics are complex and play a complicated and nuanced role in student learning and teachers' instructional practice. While there is substantial research that documents the relationship between certain teacher characteristics (years of education, expertise, subject-matter knowledge) and student achievement, there is not a similar extensive body of literature that examines the relationship between certain teaching practices and student outcomes in mathematics. Based on their review of the literature, however, Hiebert and Grouws suggest that two types of opportunities to learn are integral to student achievement: opportunities to develop skill efficiency (which has also been called "procedural fluency") and opportunities to develop conceptual understanding. They find, based on review of carefully selected studies, that the practices that promote procedural fluency include explicit teacher-directed questioning, instruction that is rapidly paced, and lots of "error-free" practice. Practices that promote conceptual understanding comprise two features: teachers and students attend explicitly to concepts—that is, "to connections among mathematical facts, procedures, and ideas" (p. 383)—and students have the opportunity to "struggle with important mathematics," meaning that they "expend effort to make sense of mathematics, to figure something out that is not immediately apparent" (p. 387). There is not much literature that attends to both procedural fluency and conceptual understanding within the same study, although some researchers have suggested that certain reform-oriented practices do promote both types of student learning. Others have suggested that these practices might be more likely to promote conceptual understanding but not procedural fluency.

As Grant (2002) describes, the teaching-learning paradigm in urban schools as depicted in film and television often presents an unrealistic picture of teaching. In addition to painting classroom teachers as heroes who triumph (usually) in tough classrooms against all odds, these films often reduce the work of teaching to inspirational chants, scripted learning, and few opportunities for students to engage in critical thinking and problem solving, even when they are learning high-level content as in *Stand and Deliver*. Grant (2002) goes so far as to say that "successful teaching in these films [*187, Dangerous Minds,* and *Stand and Deliver*] is portrayed as the receiving of love and adoration from students, not in the improved quality of students' lives as a result of their education" (p. 85).

The literature on effective teaching practices suggests, in contrast to the platitudes about and simplistic view of teaching often portrayed in film and television, that effective teachers hold high expectations for students; expose them to and ensure that they learn rigorous content; and connect their real-world, out-of-classroom, or cultural experiences to classroom learning. Gloria Ladson-Billings (1997), for example, describes effective teachers of African American students as those who enact culturally relevant pedagogy. In her study, Ladson-Billings found, in part, that effective teachers believed in the competence of their students, maintained fluid student–teacher relationships and encouraged student collaboration, and were passionate about knowledge and learning. In mathematics, Gutstein, Lipman, Hernandez, and de los Reyes (1997), Leonard (2008), and others have described how effective culturally relevant pedagogy practices might be applied in mathematics classrooms. Often, these approaches are described in tandem with reform-based instructional practices. Gutstein and colleagues (1997), for example, point out teachers' efforts to build meaningfully on Latino/a students' cultures in designing real-world mathematics problems and in facilitating class discussions.

However, much of this research does not take into account how *students* evaluate teachers, and how they describe teachers who inspire them, motivate them, and teach them effectively. A study by Corbett and Wilson (2002) found that "inner-city, low-income middle and high school" students described good teachers as those who "made sure that students did their work, controlled the classroom, were willing to help students whenever and however the students wanted help, explained assignments and content clearly, varied the classroom routine, and took the time to get to know the students and their circumstances" (p. 18). These sentiments were echoed by students at Lowell High, when they reflected on their best teachers' instructional practices. However, the students in Corbett and Wilson's study were also were adept at understanding that teachers of multiple styles and personalities could exhibit these positive qualities vis-à-vis teaching. In short, for good teachers, "demeanor, sense of humor, and charisma—as well as any other personal characteristics—were unimportant" (p. 18). Students were not expecting teachers to be their friends, or to entertain them all the time. These findings were echoed in the work of Delpit (1995) and others who found that students highly valued firm, fair teachers who had high expectations.

For teachers who are not effective, students report various ways in which teachers make their disinterest apparent: The level of classroom discipline, academic expectations, challenging work, and caring are all indicators for students of teachers' interest in them as students and human beings (Corbett & Wilson, 2002; Delpit, 1995). For example, the Black male students in Polite's (1994) study characterized the majority of their teachers and counselors at their high school as "uncaring" because of their perceived disinterest in students' coursework. As with the students in Polite's study, Ferguson (2002) found that Black students placed a greater premium than did students from other groups on the role of teacher encouragement in their academic progress. Lowell students also pointed out that teachers who were "standoffish" or not good at classroom management impeded their learning and that of other students.

As the students in Corbett and Wilson's (2002) study note, students value teachers who make the effort to establish relationships with them. Several researchers—Delpit (1995), Ladson-Billings (1995), Noddings (1995), and Valenzuela (1999), for example—have described components of teacher-student relationships that foster student development and learning. An ethic of "caring" is important to students, but this construct is a complex one. For example, as Ladson-Billings (1995) notes in her study of seven effective teachers (4 Black, 3 White) of African American students:

> The teachers were not all demonstrative and affectionate toward the students. Instead their common thread of caring was their concern for the implications their work had on their students' lives, the welfare of the community, and unjust social arrangements. Thus, rather than the idiosyncratic caring for individual students (for whom they did seem to care), the teachers spoke of the import of their work for preparing students for confronting inequitable and undemocratic social structures (p. 474).

Rodriguez (2008) draws on international literature and his own research with students in urban, high-poverty schools to describe the importance of "recognition," which "signals to the student that the teacher knows, cares [about], and values the student" (p. 440). In urban schools, in particular, the frequent characterization of urban students as the "other" may inhibit recognition. The students in Rodriguez's study and others talk very powerfully about the importance of teachers' knowing them, valuing them, and encouraging them to have high aspirations. Teachers who exhibit these qualities, like those in Delpits's, Corbett and Wilson's, Ladson-Billings's, and others' research, do not accept failure from students.

AT LOWELL HIGH: STUDENTS' PERCEPTIONS OF MATHEMATICS TEACHERS AND THEIR INSTRUCTIONAL PRACTICES

In Chapter 1, we pointed out that there were some key gender and ethnic differences in how students thought about their teachers, and their teachers' perceptions of them. Overall, however, Lowell students agreed (on a scale of 1 to 5, with 1 = strongly disagree, 3 = neutral, 5 = strongly agree) that their current math teacher "cares about how well you do in his/her class" (4.35) and were a bit more neutral, but still positive about whether their teacher thinks they're smart (3.73), cares about you as a person (3.75), or thinks that you are good in math (3.37). Lowell students felt that it was unlikely that their teacher would recommend them for an honors or advanced mathematics course (2.61), but this may be because there are few honors or advanced courses at Lowell and the average student grade in mathematics is C.

In the remaining portion of this chapter, I describe qualities of good mathematics teachers, as described by Lowell High students. (I discuss these issues more, and what students find ineffective, in Chapter 4). Lowell students, like other students in the research literature, describe their best teachers (from elementary through secondary school) as those who develop strong interpersonal relationships with them, have high expectations for their learning, have a strong command of their classrooms, and have a strong knowledge of and enthusiasm for their subject matter.

> I was really hanging out with the bad crowd [in 8th grade]. [My teacher] would come and talk to me, like, "I know you can do much better than this," and she just made math a good learning experience.

> Some teachers recognized my talent and motivated me to do better.

Lowell students reported that it was very important to them that teachers "respect them" and show concern. They felt it was important that teachers not treat teaching as though it were "just a job."

Twelve of the 21 high-achieving students interviewed mentioned explicitly that their success in mathematics was due to the fact that many of their past and current Lowell teachers were good teachers. In particular, they liked "how [Lowell] teachers explained things." Noted Gabriela, "I love math, and the teacher makes it fun in the way he explains everything with details." Alicia mentioned that her Lowell teachers "don't just tell me the answer, when I have a question. They stick with me and they challenge me, and I like to be challenged. There's always something new to do in math class." Several students mentioned other aspects of the teacher-student relationship, including the importance of feeling that the teacher cared about them (Ladson-Billings, 1997; Valenzuela, 1999) and how this caring contributed to their achievement. Yvette wrote on her influences map (an example of this map appears in Appendix B) that "the school is very helpful as well as the caring teachers. They take the time to be patient and teach. If it weren't for them, I wouldn't know math!"

Students believed that their previous teachers were an important part of students' mathematics success and how they viewed themselves in terms of their mathematics ability:

> They [former math teachers] always told me that I was really good in math. (Nicholas)

> Mr. Lopez . . . the way he taught, made you interested in math. He just made me feel that, not to give up, that you can do it. That's the only way to solving problems. (Naomi)

> Like my 6th-grade teacher was a pretty good teacher . . . the way he explained things and then sometimes he would make us do the problem together, and sometimes alone. (Louis)

Sometimes the student's teachers' expectations were greater than those of the student:

> The teachers recommended me [for an advanced class] 'cause we were supposed to take a test, but I didn't think to myself that I was up there, so I didn't take the test. (John)

When students questioned their own competence in academic matters or lacked knowledge about options open to them, current teachers and counselors helped them to make important academic choices:

> [The guidance counselor] said it would look better when I'm applying for college [to take an advanced class]. (Lamont)

> They asked if the 9th-grade math was a little too slow, and I said yes, and they gave me the test and I passed it so they moved me up into the 10th grade [math]. (Yvette)

> He knew I was bored and I already knew everything he was covering. So he recommended to a guidance counselor, to put me in the higher math. (Joy)

Lowell teachers were attentive and aware of students who exhibited potential for high mathematics achievement. Teachers contributed to the intellectual milieu of students in ways that extended beyond their classrooms:

> Some of my friends ask to study with me, especially now since Mr. Quigley [her mathematics teacher] announced, "Yvette has the highest grade!" (Yvette)

> He [his math teacher] challenges me; he gives me puzzles, math puzzles to do. Like, to exercise my mind or something. *[Laughs]* (Ian)

One Lowell teacher in particular, Mr. Lewis, was well regarded by his students. Several students describe Mr. Lewis's work:

> Mr. Lewis, he's really encouraging, and he doesn't seem like he's just a teacher, he's just there to get paid. He seems like he wants you to learn, and if not, he's gonna go over it and over and over it.

> I like when Mr. Lewis inspires me a lot. . . . He'll be like, "Oh, well, you got this on a test, and we know you can do better, but this was great."

Linus had this to say about Mr. Lewis:

> He knows how to communicate with students good. Like he's a great teacher, even after school, it's like the students and him, we play basketball after school. You usually don't hear of students and teachers hanging out. . . . He talks about interesting facts that have to do with math. Like, for example, the other week, he told us something about soda . . . and they did a test on kids who drink soda, and like how much grams of sugar it has and how much weight they gained every week. And that is really interesting.

Further, Linus wrote on his influences map that "Mr. Lewis—he's always encouraging me to be even smarter."

Despite the lack of student reports about established contact between the home (in particular, their parents) and the school, students noted that school adults at Lowell intervened in course placement, encouraged them to do homework and classwork, provided enriching mathematics work on occasion, and were good teachers. Students also described additional school policies at Lowell that they viewed as helpful to their success in mathematics. For example, three students mentioned that the school provides free tutoring to help students pass the Regents mathematics examination, necessary to earn the prestigious Regents-endorsed diploma. One student mentioned that the small size of the school, and thus the math classes, made it easier for teachers to help students individually. In addition, as Linus describes, there are ways in which teachers in classrooms strive to make mathematics engaging for students. I describe some of these activities and practices in the next chapter.

CHAPTER 4

Engaging Urban Students' Mathematical Interests to Promote Learning and Achievement

As part of a project spearheaded by two humanities teachers (English and social studies) and a journalist, 14 urban high school students recasted the SAT as a test measuring students' intelligence in urban contexts. Naming the test the "SAT Bronx," the participants wanted to make the point that the SAT as an examination depends quite a bit on cultural knowledge that is considered "mainstream." The larger goal of the project, set up as a school club for interested students, was to highlight "what urban students already know and can do" (What Kids Can Do, 2008, p. 3). Items on the SAT Bronx were multiple choice and involved numerous vocabulary and reading comprehension items, alluding to students' multiple understandings, vocabularies, and experiences in urban youth contexts. The supposition was that adults (or other teenagers) who had not shared these experiences would not do well on such a test without background knowledge that the student writers provided. The mathematics section, "How Do You Get There?," involved "cost-benefit analyses" related to various transportation situations involving the New York City subway and students' trips to school and to work.

The writers of the book and test (students, teachers, and a journalist) argue in the text that while the SAT is important, it is rooted in a history and structure that is designed to exclude certain groups of students. As one student who helped to design the SAT Bronx said:

> We wanted to turn the tables around, and actually have situations and experiences on the test that dealt with us. If you have taken the SATs, you know there are things like *trigonometry and vocabulary that half of us don't even deal with* [my emphasis]. We were kind of sick of it! We wanted to have something that related more to us, that was from our hearts. (quoted in Cushman, 2009, p. 185).

In general, the work of the students in the SAT Bronx project, and the student's quotation above, that trigonometry is something that "half of us don't even deal with," is underscored by the contention of numerous observers that the processes of doing school mathematics bear little resemblance to the processes by which mathematicians do mathematics and hold little interest for students. Schoenfeld's (1985, 1988) seminal studies reveal that students believe "formal mathematics has little or nothing to do with real thinking or problem solving, that mathematics problems are always solved in less than 10 minutes if they are solved at all, and that only geniuses are capable of discovering or creating mathematics" (Cirillo & Herbel-Eisenmann, 2011, p. 69). These beliefs relate to students' lack of interest in the underlying understanding of algorithms and procedures (and also, teachers' practices that reinforce that understanding these algorithms and procedures is unimportant); they believe that these have been disseminated "from above" and that mathematics has no relevance to their contemporary lives.

Unfortunately, despite the possibilities for students to be engaged in mathematics for its aesthetic qualities, and the crucial importance of mathematics in science, statistics, technology, and the like, students are often largely unaware of these issues. In particular, high school math students often find the mathematics classroom disengaging or uninteresting (Boaler, 2002), and as stated in Chapter 1, there is national evidence that young people become less interested in mathematics as they progress through school. In fact, students' positive attitudes toward mathematics are primarily centered around it being a "necessary" evil to handle money, to pass examinations, and to gain access to college.

Recent reforms in mathematics education have suggested that mathematics should be "meaningful" to students. However, it should be noted that occasionally the development of meaningful problems and activities, despite their efforts to be relevant to students, are problematic in their content, scope, or format. Content-wise, the mathematics in the problems may not be appropriately linked to student abilities—problems may be too easy (focusing on rudimentary content too basic for 10th-graders, for example) or too inaccessible. In terms of scope, problems or activities may be too time consuming for teachers to use in the classroom. This is a key reason why teachers report that they do not use reform-based activities more often in their classrooms. Finally, the format of the problems or activities themselves may present real obstacles to students' learning and understanding. For example, researchers have found that the "wordiness" of word problems or project-based activities have caused students of limited English proficiency or students who lack grade-level literacy to struggle with problems, even when their mathematics abilities are on or above grade level.

It is also important to consider how students' everyday activities and practices lend themselves to learning and engaging with mathematics. Researchers who explore out-of-school mathematics practices and behaviors, such as Masingila, Muthwii, and Kimani (2011), find that "students gain mathematical and scientific power when their in-school mathematics and science experiences build on and formalize their knowledge gained in out of school situations and when their out-of-school mathematical and scientific experiences apply and concretize their knowledge gained in the classroom" (p. 89).

How, then, should we ensure that mathematics is an engaging subject for young people? Before considering how to develop strategies, activities, and experiences that are engaging for high school students, let me describe what I mean by engagement. Like many researchers, I suggest that engagement should be considered as a construct that simultaneously encompasses behavior, emotion, and cognition (Fredricks, Blumenfeld, & Paris, 2004). Although there is much research that considers these three components separately, exploring engagement as a multifaceted construct "may provide a richer characterization of children" (p. 60). Considering these three aspects as an integrated set of factors has guided my work with high school students in mathematics. As Fredricks, Blumenfeld, and Paris (2004) describe:

> Behavioral engagement draws on the idea of participation; it includes involvement in academic and social or extracurricular activities and is considered crucial for achieving positive academic outcomes and preventing dropping out. Emotional engagement encompasses positive and negative reactions to teachers, classmates, academics, and school and is presumed to create ties to an institution and influence willingness to do the work. Finally, cognitive engagement draws on the idea of investment; it incorporates thoughtfulness and willingness to exert the effort necessary to comprehend complex ideas and master difficult skills (p. 60).

Considering these three components of engagement simultaneously allows us to deeply understand young people's attitudes and actions around mathematics. For example, students' behaviors in mathematics class may be a function of their beliefs about mathematics and school, their relationships with the teacher and their other classmates, and their motivational stance on whether mathematics is worth learning as a discipline on its own, or rather just to get a good grade or pass tests.

Research has demonstrated that student motivation to learn and achieve is strongly related to students' attitudes toward content areas, school, education, and learning; their performance in school and on achievement tests; other affective constructs like confidence, engagement, and interest; and school and classroom behaviors. In addition, there is substantial evidence that classroom instruction and practices affect students' attitudes, performance, and behaviors, as described in Chapter 3. What is less clear is how these relationships are mediated by other factors, and how these relationships are interrelated. For example, some studies have shown that teachers' high standards for mathematics learning, especially in reform contexts, can, at least initially, increase student mathematics anxiety, which in turn can have a dampening effect on student performance. But this effect might be mediated by continued exposure to these high expectations, because arguably a different type of expectation for student learning contradicts how students have traditionally learned mathematics. (Parents are not immune to this phenomenon either—Lubienski's work [2004] shows that when choosing traditional or reform-based mathematics courses for their children, parents were more comfortable with traditional courses). Other studies have found that teachers' inaccurate assessments of students' motivational issues (Givvin, Stipek, Salmon, & MacGyvers, 2001; Fennema, 1990) can lead them to provide faulty interventions or instructional practices to these students. Much more research, especially longitudinal studies that trace students' performance through critical stages in school mathematics, on these issues should be conducted.

All this research suggests that it is not enough that students are exposed to rich and rigorous mathematics content, but because of all the myths and social issues related to math there are important socialization issues that must be acknowledged. The National Mathematics Advisory Panel (2008), National Research Council (2001), NCTM (2000), and others have all suggested that dispositional factors are important in developing students' competence in mathematics and should be attended to in mathematics activities. Specifically, Zollman, Smith, and Reisdorf (2011) argue that the development of students' mathematics identity, and the activities through which teachers facilitate it, are a critical component for learning in mathematics. They suggest that teachers provide learning opportunities in class in which students' self-determination and self-regulation are facilitated, but that also focus on social aspects such as cooperative and collaborative goals as well as creating an engaging classroom environment.

WHAT ACTUALLY HAPPENS IN SCHOOL MATHEMATICS CLASSROOMS

Despite mathematics education reform initiatives driven by the National Council of Teachers of Mathematics (NCTM), the National Science Foundation (NSF), and the National Research Council (NRC), among other groups, mathematics in the United States is still taught in a largely didactic, teacher-centered way (McKinney et al., 2009). Indeed,

> most American teachers have a conception of mathematics as a static body of knowledge, involving a set of rules and procedures that are applied to yield one right answer. "Knowing" mathematics means being skillful and efficient in performing procedures and manipulating symbols without necessarily understanding what they represent (Thompson, 1992). (Stipek, Givvin, Salmon, & MacGyvers, 2001, p. 214)

Even a study of NCTM members (McKinney et al. 2009) found that those members tended to use lectures, drill and practice, and teacher-dominant instruction very frequently when compared with "alternative" approaches like using manipulatives and providing hands-on and problem-based learning activities. School mathematics thus is focused on rote learning, is primarily teacher centered, and is not engaging for students. International studies such as the Trends in Mathematics and Science Study (TIMSS), which has included video studies, showed that most mathematics classrooms in the United States do not actively engage students in mathematics learning.

Engagement in mathematics education research has been measured in several ways: surveys of students or interviews of students about their attitudes toward mathematics and school, observational methods that focus on teacher and student discourse in mathematics classrooms through real-time observations or video clips, event history analysis to capture the duration of students' attention during particular classroom episodes, and sampling methods that involve recording students' behaviors and affect during random intervals during classroom activities (Uekawa, Borman, & Lee, 2007). These studies have found, generally, that students tend to have higher engagement levels (including measures of confidence and interest) when they find the mathematics classroom activities to be relevant, challenging, and academically demanding and believe that their teachers have high expectations for them (Woolley, Strutchens, Gilbert, & Martin, 2010).

There is also evidence that students' cultural backgrounds are related to how they perceive the mathematics classroom. Using the Experience Sampling Method (ESM) with a convenience sample of 345 students in four different cities, Uekawa et al. (2007) found that students belonging to different cultural groups and in different regions of the country may find different classroom activities engaging: lectures, group work, and individual seatwork contributed to different levels of engagement for Latino/a students in Chicago and Miami sites, but not in El Paso, for example. Asian students regardless of region had higher levels of engagement when the mathematics classroom activities were not cooperative. Initially, this appears to contrast with Fullilove and Treisman's (1990) finding that Asian and Asian American college students spent a great deal of time in cooperative learning groups outside of the mathematics classroom. Some students may feel that in mathematics class individual seatwork and lecture are the appropriate forms of activity, and outside of class, when one can choose one's own partners, is when one engages in cooperative learning activities. Overall, Uekawa and colleagues found that "white students in general and Latino students in El Paso strongly preferred lecture to individual seatwork," while "Black students, who were generally higher [than other groups] in engagement levels were little affected by differences in types of classroom activities" (p. 36).

Many researchers have found that classroom environment and social dynamics (between the teacher and students, especially) have a critical effect on academic outcomes for Black students in particular. However, researchers caution that pedagogical strategies most depend on local and classroom contexts and that teacher and student dynamics should not be based on essentialist views of students' race and ethnic group membership. It is a fine balance, because some "scholars [have] warn[ed] of detrimental effects that reform-initiated practices that are uncritically embraced and carelessly implemented may have on students of color (Delpit, 1988) and students from low socioeconomic conditions (Lubienski, 1996)" (Cahnmann & Remillard, 2002, p. 181).

Martin (2009b) cautions that taking reform practices wholesale without considering students' contexts and interests and specific cultural experiences might end up inadvertently exacerbating disparities in performance, as noted by Secada (1992).

Many mathematics education reform initiatives since the 1980s have highlighted the need for more active engagement of students, and more "student-centered" classrooms. As Stipek et al. (2001) note:

> Inquiry oriented mathematics educators take a more dynamic view of mathematics, conceptualizing it as a discipline that is continually undergoing change and revision (Prawat, 1992b). . . . They recommend classroom practices that actively engage students . . . in activities that require reasoning and creativity, gathering and applying information, discovering, and communicating ideas. (p. 214)

However, as mentioned earlier, reform efforts often challenge teachers', students', and parents' notions of what mathematics is and how it should be taught. In addition, some researchers suggest that reform mathematics has the potential to improve student engagement and achievement, but these studies largely have not compared levels of student engagement and achievement in conventional mathematics classrooms with those in reform classrooms. Indeed, the National Mathematics Advisory Panel (2008) pointed out the lack of quality long-term research in these areas of mathematics education. Another issue is that many standardized assessments used in the current era of accountability have generally defaulted to assessing students' basic skills and procedural fluency, rather than students' problem-solving abilities and conceptual understanding. In large part, this is due to requirements for frequent assessments as well as the reality that it is more expensive to assess student performance using something other than standardized, multiple-choice assessments. While some have suggested that teachers may be spending more time focusing on rote activities and basic skill drills because of the accountability climate, the evidence is mixed. Vogler and Burton (2010), in a study exploring teachers' instructional practices as a response to high-stakes testing environments, found that teachers use a balance of standards-based and traditional practices and tools in the classroom. Others suggest that teachers predominantly use traditional practices with a smattering of reform-based practices.

Stipek and colleagues (2001) found that teachers in their study who held more traditional views of mathematics and learning "had lower self-confidence and enjoyed mathematics less than teachers who held more inquiry-oriented views" (p. 223), conjecturing that these teachers are more comfortable with beliefs and practices that require less expertise in content and less judgment in classroom in-the-moment instructional decisions. These teachers, then, might feel much more comfortable with a prescribed curriculum in which the textbook is central to the mathematics being taught and learned, while inquiry-oriented teachers are more comfortable with the open-ended discussions and questions that students might have (and find more engaging) in inquiry-focused classrooms. Gutiérrez (1999) noted in her study of effective teachers of Latino/a students that these teachers were less drawn to the textbook, developing materials for students that facilitated discussion and problem solving. But these teachers also exhibited considerable expertise.

Further, it should be clear from the statements of students in Corbett and Wilson's (2002) study described in Chapter 3 that "holding students' attention and making them happy" (Grant, 2002, p. 83) is not the primary purpose of teaching. Popular films about teaching in urban schools leave the impression that if these two things are accomplished, then "learning will be effortless and automatic" (p. 84). This does a real disservice to novice teachers, who may come away from these films with this message. The voices of students reveal that they do not see engagement solely as being entertained in school.

NONENGAGING MATHEMATICS EXPERIENCES: LOWELL HIGH STUDENTS' VOICES

Sadly, an overwhelming number of reports from students describe the mathematics classroom as a place where they learn that mathematics is a set of routines and procedures that are important for test-taking. Much of what students describe as problematic about the mathematics classroom has a lot to do with the practices of their teachers and the curriculum to which they are exposed. Typically, Lowell High School students described their nonengaging mathematics classrooms in the same way. They are rooms in which students are "too scared to want to raise your hand to ask for help, so you try to do it yourself, and then you get a bad grade." They are rooms in which the class is "boring," and the teacher "just write[s] something on the board, explains it once, and if you don't understand it, then that's it":

Q: Was the teacher responsive if kids had questions? Did anybody ask him questions?
A: No, because he would just write the lesson, explain it once, and then he would sit down. And in the class, that's it.

One time, a student asked her [the teacher] to explain something because the students did not understand the problem. Instead of showing the steps at a slower pace and explaining the problem in more detail, she just goes through the problem like she first explained it.

He expected us to read and catch on, instead we just guessed most of the questions. He didn't spend time to show how it worked. We grew used to that. Working on math problems by ourselves.

We had to copy down notes for like the whole period, and we didn't even use them.

Much of high school students' descriptions tend to focus on the affective aspects of the mathematics classroom—it is boring and repetitive; they feel scared. When adults reflect on their adolescent experiences in mathematics classrooms, they also point out that the math curriculum over time is repetitive (Schmidt, Houang, & Cogan, 2002) and consists largely of rote work. As one mathematician noted,

> School math wasn't that interesting. Because I think coming up through the years everything was just plug and chug. I'm not sure if that was all that interesting to me, all that plugging and chugging.

ENGAGING MATHEMATICS EXPERIENCES: VOICES OF LOWELL HIGH STUDENTS AND MATHEMATICIANS

In interviews with Lowell High students, they, fortunately, also describe engaging experiences with mathematics in school, focusing on mathematics curriculum and teachers' practices. These experiences are not limited to their high school experiences but also encompass earlier school experiences. In addition, on occasion students describe in-school experiences that overlap with their out-of-school mathematics interests. For example, some students described their interest in mathematical games and puzzles:

> I like a lot of math games, like when I was young, I used to like 24. . . . It's a game where there's a bunch of cards, and in the middle . . . hmm. I forgot how it went. But there's a bunch of operations on it and there's different numbers on it, and you have to try and figure out how to use the numbers and the operations to get to 24.

> I learned, like, the way he teaches math is different from the other teacher but it's fun. Like we do riddles in class.

> Yeah, Mr. Lewis is cool, you know. We get along or whatever, we like talking about math, and like he challenges me, like he gives me puzzles, math puzzles to do . . . like to exercise my mind or something. [Laughs] It helps, like you know, it made me think fast or whatever.

It's clear that engaging activities are not necessarily "stand-alone" but are strongly tied to students' assessments of teachers' pedagogical strengths:

> He was really vibrant. Like . . . let me give you different examples. Like everything wasn't just math, like numbers. He would give you more lively examples, like, let's say you went to a movie . . .

They gave me notes that made the class interesting. And also gave some good examples on the board.

Q: Why was this person your favorite math teacher?
A: She always explained things. . . . If someone was like stuck or didn't understand she always explained again and helped. And we were always, like, linking things to real life.
Q: Can you give me an example?
A: Oh, yeah. This is like a perfect example: She was teaching us about area and like how to find the area of a rectangle and she just filled it in, and we used blocks to find the area.

She was a good [9th-grade] math teacher because she would make it fun for students. She would give examples and give out, um, figures and shapes for us to understand and she made jokes and made up stories. . . . And sometimes she would ask us to participate.

My favorite teacher was Mr. Solomon, in 8th grade. He used to make songs so we could remember what we were doing. And I think if it wasn't for him, I wouldn't be doing good now in math. 'Cause he taught us ahead of what we were supposed to do.
[Q: What was the song?]
   It was to get the y alone. He had a song, he had a song and a dance to it. And he really imprinted to us that we had to have the y alone. He was so funny. It was so funny. . . . I still remember that, I don't think I'll ever forget that.

Well, my favorite math teacher makes things like funny. He makes it really interesting to learn, and he gives examples and everything. He's teaching each one, every single one of them, he looks at you, and he tries to explain, do you understand, and he asks if you do understand and you can! It's easy to, he can teach you really easily.

Like, he's really encouraging. He doesn't seem like he's just a teacher, he's just there to get paid. He seems like he wants you to learn and if not, he's gonna go over it and over it and over it, so.

Just every day there was something new to do. Mr. Lewis always had something new to do, we didn't do the same stuff every day.

Both male and female Lowell students described teachers' abilities to engage them in mathematics similarly, although boys were more likely to mention games or television shows as memorable experiences that they connect with mathematics teaching and learning.

When talking to adults who are mathematicians about mathematics experiences they found engaging during adolescence, they report school-based experiences as well as out-of-school experiences (Walker, 2009b). It may be that the Lowell students might reflect differently about out-of-school mathematics experiences as they grow older or if they pursue a mathematics-based career. For many of the mathematicians I have interviewed, these experiences were the spark that made them see mathematics as something that was not limited to textbook exercises but also a discipline that existed outside of school, and in interesting ways.

For example, this mathematician's experience growing up in a rural area in the late 1950s still resonates for him today:

> One thing I remember is when I was in about the 9th grade, my uncle worked for a construction company. He saw the foreman using a slide rule. He just got curious about it, so the foreman said, "Well, next time I place an order for equipment, I will order you one if you'd like." So the slide rule came with a thick manual about trigonometric functions and those such things. . . . In the family, people thought that I was some sort of bookworm because I was always reading books. So he just gave it to me. I wanted to get to the basics of the thing. I wanted to understand it, so I actually read the manual. I knew enough algebra and trigonometry to figure out most of the scales. For me it became a hobby.
>
> Now, I graduated high school in 1960. I got this slide rule maybe in 1957 or somewhere in there and learned about trigonometry. When I was supposed to graduate, a good buddy of mine and I were idling time away walking down a country road headed home. We came upon a man who was surveying some land. He needed two strong fellows to help him pull some chains. He told us that he would pay us $.75 per hour to do this. This is a lot more than you could make working on the farm. You could earn two or three dollars a day by working on the farm, but here is a guy who is going to pay $.75 per hour. I thought that this was an enormous sum of money to pull these chains. . . . When this guy started talking to us, he had a transit. He would set it up and sight through here and swing around through a certain angle and sight through there. He could compute the distance between two faraway points. When he found out that I knew a little trigonometry, he started teaching me how to use this transit. He was so impressed with me and I was so amazed by how much money you could make using this trigonometry. So I said right away that I wanted to be an engineer because I thought that engineers made even more money than high school math teachers. I wanted to be a civil engineer.
>
> So this was a moving experience. I wish that students at the 10th-grade level could see something like this, where "here is something I am learning in school that is being used to earn money." Meeting that engineer who was

surveying land . . . he was friendly enough to teach me things about how he was actually measuring the distance, and in doing this without having to jump across that ditch over there to get to [there]. Now we had studied about triangles and all—if you know this side and you know this side and you know the angle between you can get the length of the third side and all that. But here it was in action. This was very powerful.

This example shows how vivid and resonant a "happenstance" out-of-school experience with mathematics can be. Upon reflection as an adult, this mathematician describes the importance of the experience in facilitating his interest in using mathematics in a career and understanding that it was related to his schoolwork in geometry and trigonometry.

Similarly, another mathematician relates a story that demonstrates for him, years later, that an experience in childhood can be useful in advanced mathematics:

My grandfather lived right around the corner from here [the college where he is now a professor]. I remember he would always have these mental challenges that he would give me all the time. . . . I actually use one of them in particular [when I'm teaching]. We were on the front porch and he was asking me—he was saying, if he walked halfway to the end of the porch, and then halfway again, and then halfway again, and so on, how many steps would it take him to reach the end of the porch? And so, I may have guessed 5 or something, I don't know. So then he actually proceeded to do it, halfway, and then halfway, and then halfway, but the idea was that he was converging—he didn't use the term *convergence*, but he never actually reached it—but he got closer and closer and closer, and of course he didn't say within epsilon. . . .

But anyway, I have fun when I'm teaching about convergence to really tap into it at this early level. One, just because I have fun telling the story—but also to give my students an idea of the sorts of things they can do with their students, because some of them may go on to become teachers, or just with their grandchildren one day, whatever the case may be. These are the sorts of things that can really bring high-level things in very early and just challenge the mind and make you think.

Another mathematician describes an experience that fostered his mathematics interest but that was not directly related to a particular problem. In fact, he was first exposed to the notion that mathematics was *not* just a tool kit—that it held interest in and of itself:

I do remember one instance—I can't remember if I was in 11th or 12th grade—but I went down to Carnegie Mellon, and I don't know how I even knew about this, but a professor from England gave a talk to high school students. I don't know. There were probably 100 people there in a large lecture hall, and it was really an eye-opener for me as a teenager. He simply talked about the open questions in mathematics, which no one had ever talked to me about. . . . It was just an important experience, hearing about mathematics, about unexplored territory. Apparently, new questions were arising every day, as opposed to mathematics just being a tool kit.

I remember that and I still talk about it. I've been saying when talking to kids about mathematics that research—there's just a lot we don't know and [the fun] of doing it for its own sake.

Our "plug and chug" mathematician from earlier in this chapter describes the influence of a teacher who made it clear that mathematics was more than just plugging and chugging:

In 6th grade I had a fabulous math teacher. . . . Everybody loved her, everybody knew that she was just the best. And she started each class with a puzzle. There was always these puzzle things going on. . . . .I'm getting teary eyed, my goodness. We just started puzzles. . . . I just remember being excited about going to her class and thinking through these puzzles and I remember her giving a puzzle and feeling like I had no clue how to answer that. I had no clue how to even begin to answer it. So it was really exciting, maybe trying to figure it out, or hearing other peoples' solutions, even if I wasn't the one who figured it out, but it just sort of being, this sort of process, not this cut and dry "yes, no, you're right, you're wrong" kind of thing . . . : It certainly wasn't that I was conscious of it at the time. I think I don't remember ever having trouble with math, but when I look back on it, it's just when I look back on it, that I feel like that gave me the first taste of sort of proving things or thinking through ideas in a little bit of a systematic way.

Thus, high school students and mathematicians report that there are instances in school classrooms as well as out of school where their interest in mathematics has been piqued.

BRIDGING OUT-OF-SCHOOL AND IN-SCHOOL MATHEMATICS LEARNING EXPERIENCES

There are some mathematics programs targeting secondary school students that seek to expose them to different kinds of engaging mathematical experiences in order to build bridges between students' out-of-school experiences and their within-school mathematics learning. One example is the Algebra Project.

The Algebra Project, designed and developed by Bob Moses, is an example of a curriculum designed to address students' engagement and to facilitate learning and understanding of mathematics concepts. As described in Moses and Cobb (2001), the Algebra Project builds on students' experiences to foster a better understanding of algebraic concepts. Based in part on Moses's experiences teaching algebra to his own children and their friends, the Algebra Project encompasses both experiential curriculum, which in Moses's view "honors how kids think" (for example, positive and negative numbers are taught using students' experiences on the Boston subways with magnitude and direction) and a focus on high expectations for underserved students. In Moses's own words,

> Instead of asking students to memorize equations and formulas, we take students on the subway and show them, step by step, how to transform their trip into a mathematical equation. In arithmetic, the underlying metaphor that kids have for addition is piling stuff on—you have 5 apples, add 5 more, and you get 10. The underlying metaphor for subtraction is taking away—you have 5 apples, take 2 away, and you have 3 left. These metaphors are insufficient for understanding algebra. . . . In algebra, the direction you move on the number line is indicated by the number and not by the operation. The operation of subtraction in algebra is now assigned the job of representing the relative positions of two points on the number line. (quoted in Checkley, 2001, p. 10)

The Algebra Project has spawned the Young People's Project, underscoring Moses's belief that as with his work during the civil rights movement, when people are given the tools to demand a change and thus increase their own agency, that has more impact than just teaching. The Young People's Project trains "math literacy workers" who are secondary school students taught by alumni of the Algebra Project and was developed by Moses's children:

> When alumni [of the program] reached their mid-to-late 20s, some of them came back to the Algebra Project and latched on to the concept of helping middle school and high school kids become math literacy workers. . . . The younger students had to learn a section of the math well enough to feel comfortable presenting it to their peers. This was groundbreaking because there weren't any models of young black kids standing up in front of other students and talking about math. These middle and high school students were a little uncertain about how they would be received by their peers. But it was considered "cool" because the young 20-something kids from the Algebra project were at the school, encouraging them. (Checkley, 2001, p. 11)

The Algebra Project is a program that is largely school based. As Morris (2006) notes, however, "social scientists have found that special programs (e.g., precollege initiatives, summer retreats) that emphasize science, technology, engineering, and mathematics can also increase students' aptitudes for and interest in these areas" (p. 256). Many mathematicians describe formal extracurricular and summer activities and programs—some school based, others not—that they deem instrumental to their mathematics development as well as fostering their engagement and socialization in mathematics. These experiences occupy spaces "in between" within-school and out-of-school mathematics, and appear to be critical to mathematicians' early mathematics learning and socialization. A number of these were targeted specifically to ethnic minority or urban students.

For example, one mathematician describes the role of a teacher who was not a math teacher in exposing her to opportunities to do more mathematics than she was getting in school:

> And even though this man was not a mathematician, he was a chemist, he very, very strongly influenced my career as a mathematician because he was actually, in retrospect I think he's probably very good at mathematics. He had gotten a master's degree from Columbia in chemistry. . . .
>
> And he was our science club advisor. And so I had him as science club advisor in 7th and 8th grade and then he was a science teacher in the 8th and 9th grade and then homeroom teacher for me. And he is the one that actually pointed me in the direction of going to [a specialized high school for science], which I didn't even know existed. But, you know, his attitude and, you know, pretty much, I mean he effectively pretty much imprinted me whatever, much of my attitude and everything was pretty much set by him I would say.
>
> Well, Charles Wilson was a chemist and I had been interested in chemistry from when I was 9. So when I left junior high I continued I think pretty much with that interest, and he encouraged me. Actually, [the university] plays a role again. He encouraged me to, and in fact I think played a strong role in getting me enrolled in something called the Saturday Science Honors Program. . . .
>
> And this was something that was sponsored by the [university]. And what it did is every Saturday it had various courses, enrichment courses for students in New York City. And it had a program in the summer as well. And I'm trying to remember the, trying to remember the right sequence. But what happened was that after the year that I graduated from junior high, that is, left the 9th grade, and just before I went on to [the specialized high school for science], 10th grade, I was enrolled in the [university] program. And this was a program, the material that they taught was basically mathematics. Much of it centered around computer programming and a lot of, a lot of elementary, a lot of enrichment, a lot of enriched material in mathematics, nice in mathematics.

Another mathematician describes his experiences in a summer program specifically targeting ethnic minority students:

Between my junior and senior year [of high school], I actually went to a "minority engineering opportunity program," or MEOP. I went down to the University of Maryland for the summer—I forget how many weeks of the summer. We took classes, and we were exposed to a variety of classes that a freshman engineering student would take. It was run by a Black teacher, and I can't remember his name. I'll have to ask my brother again, because he always remembers these things. He was actually from Baltimore, and he was one of the founders of the program. My younger brother went through the same program at Howard University. I went through the one at the University of Maryland. You could choose which one you wanted to go to, and I wanted to do the University of Maryland.

Those experiences at Poly [High School] and those experiences during the MEOP program are the ones that are definitely the trigger points of my life. I decided that I was going to become an engineer.

Finally, a third mathematician describes the role of his principal in connecting him to a summer enrichment program:

Sometime in the spring of my sophomore year, the principal called me out of my chemistry class and told me that there was an opportunity to go to Yale for the summer. He'd apparently been traveling in New England and had heard about this program for kids who were not from the socioeconomic stratum from which Yale usually draws. Really, all the kids were drawn from east of the Mississippi, from all over the place. There were some local kids. The focus of the instruction was on mathematics, mathematics that you wouldn't normally see as a 10th grader. They taught us something about linear programming and mathematical induction and things like that, that required sophistication, but not a whole lot of background.

Another type of enrichment program becoming increasingly popular in the United States (having been developed in Eastern Europe in the 1800s) is the "math circle." Math circles are extracurricular programs with the goal of "sharing the intellectual appeal and beauty of mathematics with as large an audience as possible" (Tanton, 2006, p. 201). Math circles vary in format and may or may not include preparation for mathematics competitions. As Tanton notes, "[the] Boston Math Circle works hard to remove any sense of competition and completely disregards labels of 'talented' and "'gifted'" (p. 201). Despite differences in format, successful math circle programs, through the exploration of carefully selected and designed problems, have "hit upon ways not only to excite young students with mathematics, but also to help young folk develop the tenacity to tackle sustained challenges via consistent—and joyful—hard work" (p. 201). In addition, math circles simultaneously engage varied constituencies: young people, college professors, and secondary teachers. Much more research should be conducted on the impact of math circles on students' and teachers' learning and enjoyment of mathematics, and on student outcomes such as performance and participation in math activities throughout secondary school and college, and beyond.

A FRAMEWORK FOR STUDENT ENGAGEMENT IN MATHEMATICS

All these quotations, examples, and snapshots suggest methods for how to engage students in mathematics. In listening to students' voices—both those of high school students and those of mathematicians reflecting on their high school experiences—we can develop a framework to promote student engagement in mathematics that is not necessarily dependent on a particular curriculum. We can also learn from researchers, practitioners, and theorists to curate the most effective means of engaging students.

From the discussions in previous chapters, and the examples, quotations, and snapshots of engagement in this chapter, several themes emerge. Curriculum and instruction within mathematics classes, as well as extracurricular activities and programs that operate within or outside of schools, or both, should incorporate the following components:

- *Attention to rigor.* Students feel validated when they feel that an activity is worthwhile and meaningful, and rigorous and challenging activities, with appropriate support when needed from teachers or other participants, promote deep mathematics learning.
- *Attention to and validation of students' everyday experiences and interests.* Students are interested in mathematics, and many have positive attitudes toward it. Engaging students' everyday experiences with mathematics, or promoting their understanding of how mathematics is useful in everyday contexts, contributes to their development of mathematics knowledge, and also their mathematics identity. They feel increasingly efficacious.
- *Focus on community.* Many initiatives designed to promote student achievement in mathematics attend to the social aspects of learning in multiple ways. They recognize that for many students collaboration and discussion promote deeper understanding. Elements of culturally relevant pedagogy and reform mathematics encourage peer discourse as a critical part of teaching and learning. Indeed, some of these projects suggest that when students feel responsible for each other's learning this promotes positive classroom behaviors as well as a shared determination to achieve.
- *Out-of-school/in-school mathematics/experience connections—content and socialization.* Students are interested in mathematics, but have been exposed to a limited kind of mathematics in school and out-of-school

contexts. Building on out-of-school experiences can be very resonant for students, and can make them appreciate or think about links between different areas of mathematics rather than seeing it as a set of discrete topics.

Some examples of programs that incorporate elements of this framework include the aforementioned math circles; the Young People's Project affiliated with the Algebra Project, which trains young people as "math literacy" workers; and various other summer programs or after-school programs (some of which were described by mathematicians reflecting on their adolescence). However, funding of these programs always seems to be a bit transient and they do seem to rely on forceful, charismatic leadership. Paul Zeitz, who works with the Berkeley Math Circle, notes, "[These] programs work because of one or two people with incredible charisma making sacrifices. There is no evidence of a program that is truly self-sustaining" (quoted in Tanton, 2006, p. 202).

But although being exposed to engaging experiences in mathematics outside of mathematics classrooms is important, students should also be able to experience high-quality engaging mathematics within their local school contexts. I argue that more students should be exposed to these kinds of experiences in their own schools, and given that many enrichment programs target students who are already doing well, it is important to continue to identify mathematics talent and interest, as well as develop real mathematics competence, among all students. In addition to mathematics enrichment, it is clear that for high school students and mathematicians alike their engaging experiences across contexts also supported the deep understanding of core mathematics curriculum in school. Further, for mathematicians especially, students were exposed to expanded formal and informal academic communities that supported mathematics socialization and learning. What is also clear is that this framework is not dependent on any one mathematics curriculum or style of teaching and, indeed, in some cases is not even necessarily teacher dependent. In the next chapter I share and describe one example of how this framework might operate in practice within an urban high school.

CHAPTER 5

Developing a Peer Tutoring Collaborative

How can urban mathematics educators build on students' positive attitudes about mathematics and students' mathematical interests, existing peer communities that support mathematics engagement, and school and teacher practices that facilitate math success to improve mathematics outcomes for urban students? In the previous chapter, I discussed several programs and initiatives that incorporate these qualities. I described a framework for student engagement in mathematics that comprises attention to mathematical rigor, attention to and validation of students' experiences and interests, and a focus on community, and links out-of-school contexts with in-school work. Many of these programs recruit students from different high schools to participate in learning opportunities off site. But I am reminded of the Burke High School assistant principal who pointed out the importance of offering these opportunities in students' own schools. What might a school-based program that incorporated a framework for student engagement in mathematics look like?

In this chapter, I propose and describe a model of a school-based, student-led initiative designed to support mathematics learning, based on my work with students and teachers at Lowell High. When I discovered that high-achieving students had existing peer communities that supported their mathematics learning, and that there were particular teacher practices that inspired and motivated students, I drew on my own experiences as a high school teacher and replicated a peer tutoring model I had developed 10 years before for students in mathematics classes (my own and those of other teachers).

At its core, the model uses peer tutoring to address underachievement in mathematics. Peer tutoring has been extensively used as a driver for student achievement (for a review, see Robinson, Schofield, & Steers-Wentzell, 2005), but the model described in this chapter incorporates other participants as integral components for student learning and support. The model is three pronged: It suggests a site-based approach to building on existing student excellence in mathematics present in a local school in an effort to improve student mathematics achievement; it seeks to address the issues of lack of teacher knowledge about urban students and their mathematics understanding; and it aims to develop and deepen mathematics knowledge, confidence, and interest among high school students while creating a supportive community of learners.

The remainder of this chapter is dedicated to sharing insights about how schools can support mathematics achievement by creating a collaborative space encompassing teachers and students, but where high school students are clearly the leaders, responsible for providing help to their peers struggling with mathematics. I describe the evolution of the program and share analyses of the interactions among tutors, tutees, advisors, and teachers who participated in the peer tutoring collaborative project; the mathematical discourse within those interactions; and the hierarchical and collaborative relationships between teachers (in-service and preservice) and students that emerged over time. I also focus on preservice teachers' developing perceptions of the mathematics abilities of urban students. In addition, I present evidence drawn from interviews of teachers and students that describes their learning of mathematics content (for high school students) and pedagogical strategies (for teachers and high school students). Evidence that demonstrates students' mathematics improvement is drawn from participating students' test scores and mathematics course grades.

BEGINNINGS: STARTING THE TUTORING COLLABORATIVE

The genesis of the tutoring collaborative occurred during participant research conducted in 2003–2005 at Lowell High School, a predominantly Latino/a and Black school serving about 300 students. At the request of the school's principal, I observed all the mathematics teachers' classes and conducted two teacher professional development sessions about mathematics attitudes and holding high expectations for students. In addition, all the teachers at the school were surveyed about their own mathematics attitudes and their perceptions of their students' potential and future success, as reported in Chapters 1 and 3. At the same time, the students at the school were surveyed about their mathematics attitudes, peer groups, and future plans, as described throughout this book. I was particularly interested in understanding the factors contributing to high-achieving students' mathematics success at Lowell. Interviews were conducted with these students, who were identified as high achieving by their teachers, about the networks that supported their mathematics achievement.

When the observations, survey studies, and interviews were completed, several issues emerged: the predominant use of the traditional mathematics teaching-learning paradigm at Lowell (where teachers lecture about procedures for solving problems and students do individual work on representative problems); the heavy reliance of Lowell's students on their teachers to show them how to do mathematics problems; and the fact that Lowell teachers were unaware of the extent of some high-achieving students' mathematics work, discussions, and problem solving with their friends and classmates. Despite the high-achieving students' perspectives, I observed that in mathematics classes, most students seemed to have little confidence in mathematics, and when working individually on problems they needed constant verification of their process and reassurance from their teachers. They did not seem to be very self-sufficient, even after the teacher had explained the mathematics content.

Yet it was clear that Lowell's mathematically high-achieving students were engaged in networks encompassing fellow students, peers outside of school, and family that supported their mathematics work (see Chapter 2). Further, these students reported that the fellow students with whom they discussed mathematics work were both struggling and successful in mathematics. The survey of Lowell's students showed that they largely had positive or neutral perceptions of mathematics (65% of students) but did not report very often working with others on mathematics problems, studying for math tests, or doing mathematics outside of school (17% of students).

Seeking to build on high-achieving students' positive peer relationships around mathematics, to provide a space for both struggling and successful students to develop confidence in doing mathematics together outside of the formal classroom setting, and to help struggling students develop such relationships, I proposed to Lowell's principal that a peer tutoring program be started at Lowell. In mathematics, several authors (Fullilove & Treisman, 1990; Hilliard, 2003; Hrabowski, Maton, & Grief, 1998; Moses & Cobb, 2001) have described the importance of students' belonging to peer groups that support their mathematics learning. Making mathematics a collaborative activity rather than an individual one is particularly useful in that learning to communicate mathematical ideas; gaining insight from peers while completing problem-solving activities; and discussing mathematical reasoning, proof, and justification are important components of developing quantitative ability (Hiebert, Carpenter, Fennema, Fuson, Wearne, & Murray , 1997; NCTM, 2000; Webb & Mastergeorge, 2003). Thus, we invited high-achieving students in mathematics to tutor their struggling peers.

In addition, I wanted to expose Lowell teachers and graduate students in mathematics education to the work and discussions about mathematics in which some Lowell students were already engaged outside of the classroom. The principal was enthusiastic, and we began to recruit teachers and graduate students to participate in the tutoring collaborative and to identify potential tutors.

I recruited several graduate students to work with the high school students during the tutoring sessions as advisors, so that the graduate students could have experience in urban schools, as many in the teacher education community suggest (Groulx, 2001; Irvine, 2003; Rushton, 2004). In addition, the peer tutoring program would provide more opportunities for graduate students to develop meaningful and positive academic relationships with Lowell students, outside of the formal mathematics classroom.[2]

COMPONENTS OF THE COLLABORATIVE

Tutor Development

The peer tutors in the program in 2005–2006 were six high-achieving students (five Latino/a and one African American) who had been selected by their teachers and principal to serve as tutors for their peers who were struggling with mathematics. Three of these tutors participated in an hour-long training seminar in February 2006 that I facilitated, in which they were asked to solve problems, review high school students' solutions to examination problems in mathematics, and brainstorm about tutoring strategies (see Appendix A for some of the activities in the training packet for 2005–2006). However, as the semester progressed, three 9th-grade tutors also began to work with the collaborative. They began coming to the tutoring program when their teacher invited them to help the 9th-grade students who seemed shy about working with the older peer tutors. These students were not trained.

About the Tutors

In 2005–2006, four tutors (all female) completed information packets asking about their attitudes and feelings about mathematics, what their friends thought about school, their college plans, and whom they identify as contributing to their mathematics success. Two of the tutors were in 11th grade, and the other two were in 12th grade. Tutors, like other Lowell students the year previously, described broad academic communities—friends, family members, teachers—who supported their mathematics success. Two tutors were considerably less confident than the others about their mathematics abilities, choosing "neutral" when asked to agree or disagree with the statement "I am not very good at math." These two tutors also expressed "neutral" feelings when asked to respond to the statement "Generally, I enjoy math." The tutors all reported that their friends didn't like math, but that their friends tended to do well in it.

We suspected that tutors, being high achievers in math, probably were more likely than other typical students at Lowell to report having conversations about math or doing math together. Two of the tutors reported working with their closest friends on mathematics class work or homework "sometimes," while the other two reported that this happened rarely. The students who reported that this was a rarity also reported that they "never" studied with friends for math tests, while the other two reported that this happened "sometimes" and "all the time." None of the tutors reported any negative peer consequences for doing well in math. Two students felt that they (and generic students) would be "complimented or admired" for doing well in math, while the other two felt that most other students would either not care or not talk about it, or were unsure how their peers would feel about their success.

Tutors were asked two open-ended questions: "What do you like about math?" and "What do you dislike about math"? Responses to "What do you like about math" were the following:

> I understand math as soon as the teacher teaches it to us.
> I like factoring.
> I like some of the formulas and problems.
> I see math as a time to put your mind to work and to just worry about doing the work. I like the fact that math distracts you and keeps you busy.

Responses to "What do you dislike about math?" were the following:

> You have to remember a lot of formulas, equations, etc.
> I dislike fractions.
> Sometimes I dislike the difficulties and the hardness of it.
> Some parts like trigonometry and the part to memorize words.

These responses underscore that at Lowell, students perceive much of math to be procedural and formula and equation based. The response "I understand math as soon as the teacher teaches it to us" makes one wonder if this tutor was being challenged enough in her math class.

High School Student Participant Recruitment

Informal conversations with the peer tutors and other students were held to determine the best ways to recruit students for the program. From previous experience, I believed that it was critical that Lowell students felt ownership of the program and that it was not seen as a project of teachers or the administration. Initially, it was difficult to obtain a room after school to host the program, and the principal offered his office for the program. Although the students and I appreciated his generosity in offering to share his space, we agreed that the message sent by hosting an after-school tutoring program in the principal's office would undermine the importance of the program being seen as student run. Further, going to the principal's office after school, even for a voluntary program, would seem to many students to be a kind of punishment.

With the tutors and other students, I discussed which days and what times the program should meet to attract Lowell students. The students suggested that after school would be the best time, initially, but as the program developed they also wanted it to run during the lunch hour. I also asked the initial set of tutors for recommendations for other tutors. The tutors helped to design fliers that they thought would attract their classmates (see Appendix C). These fliers were posted in classrooms and hallways at Lowell and were distributed in class by teachers and students in smaller sizes. These procedures have continued in the multiple incarnations that the program has run at the school, as I will describe later.

Advisor Recruitment and Training

Graduate students in mathematics education, including preservice teacher candidates, volunteered to participate as "advisors" or "mentors" to the high school student tutors for 1 day each week. The advisors were told that they would serve as resources to the high school tutors if the latter needed help with a mathematics topic. If there were more tutees than the tutors could handle, advisors would help tutor as well. Nine advisors participated in the tutoring program.

Because the collaborative was designed in part to provide additional opportunities for mathematics education graduate students to engage with urban high school students as well as to contribute to their development as participant-observers and researchers, I designed a reflection and research component. Advisors were trained to take field notes and document tutoring interactions and were supervised by a lead graduate assistant who compiled the field notes. Each advisor completed a research memorandum focusing on a peer tutor–tutee dyad or a series of tutoring interactions.

Session Organization

Tutoring sessions took place after school in a mathematics teacher's classroom 3 days per week from 3:30 to 5:00 P.M. Because of a Department of Education requirement, a certified teacher was required to be present at the tutoring session. This teacher, Mr. Tate, attended most of the sessions (when he was unable to attend, another teacher did so). Tutors and advisors were required to participate in tutoring sessions at least 1 day per week. Students who sought tutoring would drop in or would be encouraged to come by their teachers. The session content usually focused on students' homework and test preparation. In the first month of the program, tutors worked with an average of 8–10 tutees each week. By the end of the program, an average of 20 tutees attended each week.

In the next sections of this chapter, I describe several findings that emerged from the work of the tutoring collaborative. These findings include evidence of students' improved performance in mathematics, where possible, but also include analyses of the dynamics of the interactions among tutors, tutees, advisors, and participating teachers; the mathematical discourse within those interactions; and the hierarchical and collaborative relationships that emerged over time. These findings come from interviews with tutees, tutors, teachers, and advisors and field notes documenting observations of the collaborative. These results reveal that the tutoring collaborative, with its emphasis on student-led tutoring, became a space for doing and teaching mathematics outside of the mathematics classroom that was endorsed by students and teachers at the school. I also describe the impact the program appeared to have on the graduate student mentors and the multiple roles the Lowell classroom teachers who participated in the program played in its development and implementation.

GENERAL IMPRESSIONS AND MATHEMATICS OUTCOMES

Based on their responses, the nine graduate student mentors strongly agreed that they were glad that they participated in the tutoring project (mean: 6.8, on a 7-point Likert scale), as reported on the questionnaire. On other items pertaining to their experience in the project, they reported that they enjoyed going to the school every week (mean: 6.3), and that they would recommend participating in this project or one like it to other Teachers College, Columbia University, students (mean 6.8). Generally, the mentors felt that their experiences would help them in their high school mathematics teaching (mean: 6.4).

The mentors were less positive, but still agreed with statements that the Lowell tutees (mean: 5.2) and tutors (5.7) had been eager to participate and that they (the mentors) had learned a lot about how students think about mathematics as a result of participating in the project (mean: 5.8). They strongly agreed that the Lowell tutees learned a lot of mathematics from the Teachers College mentors (6.2), and less strongly that the Lowell tutees had learned a lot of math from their Lowell tutors (5.7).

Several open-ended items gave mentors the opportunity to comment about their experiences further. When asked, "What was the best thing about participating in this project?" mentors often responded that they enjoyed seeing high school students work together and learn from each other:

> [The best thing was] seeing students take the initiative to get tutored/be tutored [of] their own free will; and seeing the student tutor and their tutee figure something out together (the "Ah, I get it" moment). (Vera)

> [The best thing was] working with the students, both tutors and tutees. Watching students tutor students helped me see what they have learned, how they perceive the material, and how they apply what they have learned. Also I saw how students made connections in helping to clarify another topic. (Lorraine)

> It was a good experience. I was glad to see that there were students that stayed after school without being forced to. (Amy)

Mentors were also asked to describe the worst thing about participating in the project. Some pointed out that "in the beginning [of the semester] not a lot of students were willing to participate." To address this, some Lowell teachers began to offer extra credit to high school students who attended tutoring sessions. One mentor, Candace, commented that some students were "coming for extra credit or because they're friends with the tutors instead of coming of their own free will." Others pointed out that occasionally "some of the students have been a bit disruptive or unwelcoming of our help" or "were unmotivated." These comments revealed that mentors often believed that students were coming to the program solely because of extrinsic factors (such as teachers' offering extra credit if the students participated) and that some students did not really want to be there and were disruptive as a result.

Several advisors made the observation that the mathematics work that students were given at Lowell was not as challenging as it could have been.

> I would say that they have mastered what their teachers have taught them. I wouldn't say that they knew extra. . . . I don't think they have that type of knowledge. But what's been presented to them, I think they understand fully to the point or to the extent where they can explain it in different ways because that is what I observed. But as far as having an extended knowledge base of mathematics, I don't think that was there.

Anecdotal evidence provided by tutees, tutors, advisors, and the supervising teacher throughout the 1st year of the program and after the end of the spring semester suggests that tutees' grades in mathematics improved. Certainly evidence provided in field notes and by tutors and advisors in interviews suggests that tutors' confidence in mathematics and mathematics tutoring increased as a result of their participation in the collaborative (discussed later in this chapter). Further, the advisors revealed that they had learned more about mathematics teaching as a result of participating. But test data from Lowell over time suggests that the peer tutoring program had a positive impact on students' performance on exams in general and on the Regents exams. For example, students' performance on the Regents mathematics examinations improved: More students took the Math B Regents exam in 2005–2006, with a pass rate of 40% (compared with a pass rate of 7% the previous year), and in 2009–2010 students' pass rates on the integrated algebra Regents improved to 77% (compared with 70% from the previous year).[2]

In the 2009–2010 incarnation of the peer tutoring program, data were collected on students' attendance and mathematics grades. While there were substantial problems with missing data, several findings did emerge. On average, students who attended the program for fewer days had lower achievement (averages of approximately 65 points) than those who attended the program for more days (averages of approximately 75 points). For these 73 students, attendance is positively correlated with fall and spring math grades ($r = .23$, $p < .05$). On average, students who attended the program gained approximately 8 points from participating in the program. Girls gained 12 points and boys gained 4 points, on average. But what was it about the program that contributed to these outcomes?

CHARACTERISTICS OF TUTOR–TUTEE INTERACTIONS

Initially, the tutor–tutee interactions followed the same structure of most mathematics teaching-learning episodes in the United States. They cajoled, disciplined, and admonished students about doing their homework, paying attention, and thinking about the mathematics at hand. Tutors generally worked with one student at a time and asked the supervising teacher, Mr. Tate, or an advisor for assistance if they had difficulty helping a student. Initially echoing the mathematics teaching paradigm in most high schools, tutors explicitly told students the procedure needed to answer a mathematics problem, rarely posed conceptual questions to tutees, and generally had one way of telling the students how to solve a particular problem.

> RACHEL: So what's the base?
> [Christopher writes down the base.]
> RACHEL: You want to start by writing down the formula.
> [She writes $A = \frac{1}{2}\,bh$.]
> RACHEL: So what's b?
> [Christopher answers her that the base is 8.]
> RACHEL: So fill in b in the [formula].
> Christopher writes $A = \frac{1}{2}\,bh = \frac{1}{2}\,(8 * 6.9)$.
> RACHEL: So what is 8 * 6.9?
> (Field notes, March 8)

However, as the semester progressed and tutors became more experienced, they began to use different strategies to address students' problems with mathematics. As one mentor noticed, a tutor had a consistent routine:

> They usually work through homework problems together. And she would help them understand the question, and then, if they were stuck, she'd usually refer to diagrams or pictures or any other math tools to remember the details. And then she'd usually try to work it out with them, so that it wasn't like she was *over* them. She was actually working through it with them.

It is unclear how and why tutors began to use these strategies, although it is possible that the unique opportunities provided by the collaborative discussions of tutors, advisors, and teachers exposed them to new and varied ways of explaining mathematics concepts. One advisor, Clair, noted:

> He [the tutee] couldn't do 80 * 4. The tutor didn't tell him to put it in his calculator, she said, "Remember how we did 4 * 8 and then add the zero." He

had a calculator right next to him. She didn't allow him to use the calculator. She kind of reminded him how to do it.

Tutors also used the strategy of providing simpler examples to help students learn mathematics. Although Coco's mathematics terminology is very informal, she is adept at using examples to frame questions so that they can solve problems on their own.

> Coco: What's good about this (x^2 - 4) is it's a perfect square [sic]. . . . That's what we call it. Let me give you some more examples of it. Like . . . well, only when it's negative like x^2 - 9. So x^2 - 9 = (x + 3)(x - 3).
> Coco writes down "x^2 - 36. (x)(x)."
> Coco: What would it be?
> BARBARA: So 6 times 6. x – 6, x + 6 (Field notes, March 30)

In addition, Coco used analogies and slang to explain concepts to her class-mates. One advisor, Jane, commented:

> The way Coco explained things was very much, with slang and with like inside jokes and everything. And so they were definitely having more fun than the rest of us, 'cause I was just like trying to explain the math. It was definitely more fun for them and I could see why students would want to come to peer tutoring and not want to come to like teacher tutoring. I mean definitely now as a teacher, I gotta make that work because even though it wasn't so much mathematical language all the time, I mean, it made sense what they were saying, what she [the tutor] was saying, and it was just in a language that the students responded to.

The tutors often contextualized mathematical terms in ways to help their tutees understand. In the exchange below, Coco is trying to help her tutee understand key geometrical definitions and how they are related:

> Coco: Bisector, what is that? You need to get bi into your head. You need to know it means two . . . like bisexual . . . people dating two . . . genders. . . . Two equal parts. So these two parts are the same, it can be cutting angles or lines . . . Bisector . . . it's bisecting the angle. You just look at it. Why is it an angle bisector?
> STEPHANIE: 'Cause it cuts the angles in half.
> Coco: So whenever you see things in two equal parts it's bisector. You got the angle bisector and the median . . . what's left?
> STEPHANIE: Midsegment.
> Coco: What's a midsegment? What do you remember about mid?

By and large, the use of informal language by tutors was commented on positively by advisors, when interviewed. They admired tutors' strategies and reported that they would adapt the examples, metaphors, and methods of the tutors in their own future high school teaching.

Advisors generally held positive perceptions of tutors' mathematics understanding and commented on tutors' rapport with their peers and also the development of students' problem-solving abilities.

> Yes. A great deal. Now, everybody grew, everybody. Everybody changed his or her approach to solving problems. I noticed that because . . . I would really look at what they were doing because I wanted to see what kinds of problems they were running into. And in the beginning, they had problems with this fractions and couldn't really remember what was said to them like 10 minutes earlier. And toward the latter part of the semester, I would notice that once they got to a certain point in solving a problem, like, they were just quick.

In addition, advisors noted that the tutors seemed to gain confidence and expertise as the program progressed, for example, Gail noted:

> [Over time] Rachel became more confident in helping the students and in her own knowledge and in her ability to help other students and . . . I think she just became more confident in that peer advisor but teacher-type role. In the beginning she would work with someone but she was frequently referring to us [the advisors] to make sure she was right or that she was doing the right thing. Near the end, she'd go up on the board and teach to a couple of people or she would sit with a group of people but would not need our help as much. She would not refer back to us.

In the view of the advisors, the tutor–tutee interactions over time became quite different from the standard teaching-learning interactions that occur in mathematics classrooms, particularly in the ways that tutors explained concepts to their tutees. One advisor noted:

> The tutors had a different approach from my approach. I recall [a time] when Hanaa was working with Asha, who got a problem wrong. And Hanaa asked her why did you think it was this, and Asha answered back. And then Hanaa said, "Well, what about this type of approach, you know, what if you would do it this way." So she more so tried to probe her to arrive at the answer. I know for a fact that Hanaa did that a lot, like she wouldn't just give the answer. Whereas, whenever I did it, I just tried to provide a different way to do it.

Later the advisor went on to describe further about Hanaa and Asha:

> Asha would say that she came to the tutoring session because she
> understood the material but whenever it came to test time she just couldn't
> perform well and talking to Hanaa, she just learned different ways to solving
> problems. And I just found it fascinating that a student could have that type
> of influence on another student.

CHALLENGES TO TRADITIONAL HIERARCHIES:
COLLABORATIONS AMONG PARTICIPANTS

Initially, tutors interacted with advisors only if the former were having difficulty
tutoring tutees. However, as the collaborative continued, tutors also sought help
from graduate student advisors on their own homework and interacted with them
about mathematics in general. The mentors often made decisions about how they
tutored students based on what they saw as problems the tutees were having. For
example, Jane's belief that students needed too much reinforcement led to a shift
in how she tutored:

> I would say do this problem and check your answer and if you run into a
> problem, I'll talk to you. But it's stopped them from doing a step and then
> looking at me and saying, "Is this right?" or "What do we do next, what do
> we do next?" It kinda allowed them to go from beginning to end and just
> think it through and then they knew if they were right or wrong. . . . I think
> the checking helped make them more self-sufficient and not need me or
> someone else to be like, "Yes, that's right, next step, good," and all that.

Because of the overwhelming focus she thought Lowell students had on homework, in her future teaching, Jane does not plan to focus on the procedural aspects
of mathematics learning:

> I really feel more strongly now than I did before that doing a couple of
> problems but understanding the process is better than doing 30 problems
> and just getting the right answer. . . . I felt like the couple of times that we
> [her students while student teaching] worked together and like, derived
> the volume or the area or whatever, they understood it better because
> they knew where the formulas were coming from. I felt that it worked
> better. And after Lowell, only working on homework problems, I saw that
> [students] could do a whole page and still not really get what was going on
> conceptually at all.

Despite her prior student teaching experience with deriving concepts and formulas, Jane did not incorporate these same ideas in her tutoring, although she thought it was important for students' mathematical learning for them to be exposed to and understand the concepts. She continued to focus on homework review.

On numerous occasions advisors helped to develop peer tutors' knowledge of mathematics. For example, John helped Rachel with her homework (field notes, March 1) and showed students (tutor and tutee) how to use the calculator to solve problems with complex numbers (field notes, March 23).

> Rachel is doing another problem working with exponents. John is looking over her work and confirming with nods or "yeahs." John encourages her to do another problem ("What else . . ."). John asks Rachel to consider the meaning of the fractional exponents. John: "Do you know what exponent ¼ [means]? It's really the 4th root. *[He draws the symbol.]* That's the coolest thing in math. So if I had x 1/12 . . ." Rachel draws the radical symbol and writes in the 12 as the index and x as the radicand. (Field notes, March 1)

Although John is sharing something new about mathematics with Rachel, the nature of the exchange is not that of a more knowledgeable adult "talking down" to a student. John saw the tutoring program as an opportunity to share more mathematics knowledge with students who might not have learned certain connected topics or who did not seem to recognize the beauty of mathematics:

> Whenever I work with them, whenever I work with students, I would always try to slip in something a little different that maybe was a different way of looking at something. . . . [Say] you're learning about triangles, this is something else about triangles you might not have known that you might not learn this year, but it's just a little nugget of information. They're always receptive. Also, we would, sometimes near the end of our tutoring, when kids are trying to wind up, we would start with some logic games, or different math games. The kids were interested in it, they enjoyed it. So, it was one of those things, it kinda made math fun. So the kids came in struggling with math but I hope by the end of this, they were really like, "Hey, math isn't always pocket protectors and protractors." Maybe they'll want to continue.

John and other advisors who had similar exchanges with tutors and tutees recognize that there are particular codes of mathematics that may interfere with high school students wanting to do mathematics or to show to others, especially peers, that they are capable of doing mathematics (Boaler, 2002; Cobb & Hodge, 2002). In some way, the advisors are trying to break down these codes and be inclusive, rather than using the language and behaviors of mathematics to exclude students. These mathematical interactions served to facilitate relationships between tutors and advisors that became less hierarchical as time progressed. In part, this occurred because advisors realized that how tutors worked with tutees was appropriate and useful in building their own knowledge. In short, the tutors became sites of expertise and knowledge for advisors as well.

Several advisors noted that seeing how tutors explained problems or concepts to tutees was illustrative. Clair wrote:

> The methods used by the tutors are most likely the methods that are understood better by the students and hopefully, I, in turn, can use them in the classroom.

Clair further noted:

> You don't always have to teach students from a lecturing perspective, you kind of get down to their level and work with them through problems and encourage them to use previous knowledge, problem-solving skills.

While this example reveals a perspective about teaching that is still somewhat hierarchical, it was clear that some advisors developed strong teaching-learning relationships with tutors and tutees over time. On one occasion, Maria had a "competition" with one of the tutors:

> She [the tutor] didn't really need to tutor anybody so we had a little competition on the board. . . . It was radicals, simplifying radicals. I think she was talking about it with her teacher that was there. And I said, "Well, let's have a competition," and I chose some really hard ones and it was good, you know. She beat me once. She's definitely got it. She's got the motivation. It comes kinda natural to her. She enjoys math and I guess being there. . . . It's impressive that she wanted to be there and help people.

The examples mentioned describe a shift in how the standard teacher-learner hierarchy was revised. This was not limited to advisors' interactions with tutors and tutees, as the next example will demonstrate.

The classroom teacher present at most of the tutoring sessions, Mr. Tate, had a purely administrative role at the outset—as described, the New York City Department of Education requires that students participating in after-school activities be supervised by a certified teacher. Initially, Mr. Tate referred tutees to advisors or peer tutors, rarely becoming involved in the actual tutoring. He would also recruit students, standing in the hallways asking, "Do you need help in math?" However, as the semester progressed, Mr. Tate, the tutors, and the advisors began to create a more collaborative tutoring space. The vignette below shows an example of the extent to which the tutor, tutee, and Mr. Tate all worked together. The italicized text provides some analysis of the vignette.

[Lori is the tutor working with Alberto. They are doing a homework set. The lesson deals with exponential growth and decay. Mr. Tate is in the room and he is reminding them that if the "growth factor is lower than 1 then it's not a growth factor anymore, it'll be a decay factor." Mr. Tate comes closer to them and looks at the book.]

MR. TATE: It's a decay factor because the graph will go lower and lower.

LORI: Oh!

ALBERTO: And this one?

LORI: That one is easy. Do you see an initial amount?

[*Although Mr. Tate is still present, Lori assumes responsibility for working with Alberto.* Lori is trying to help Alberto tie the exponential function in its general form ($y = a * bx$) with the exercise they are working on, so he can substitute the values and obtain the specific function.]

ALBERTO: One.

LORI: And the growth factor?

[They both try to determine what the growth factor is, but seem to have some trouble. Lori calls Mr. Tate and asks him for help. He helps them realize what the factor is very quickly. Still, for the next exercise they call Mr. Tate again. He begins by asking them what number they have to look at to determine how many places they will move the decimal point. They are analyzing the term that contains the growth/decay factor. The three of them are not quite sure what the factor turns out to be, they review the lesson in the book in search of hints or similar cases. *Here the three participants in the exchange—teacher, tutor, and tutee—are all engaged in determining the solution to the problem.*]

ALBERTO: Oh, we have to do this thing right here, 'cause it's the percent increase.

[*The tutee is the one who expresses a solution first.* He looks at his notes and he believes he finds a way to solve it. Lori reads the notes also and begins to understand as well. They go back to the exercise they are working on and try to connect it to the notes, although they can't quite find a way. Mr. Tate leaves them and checks his own book. After some minutes, he comes back and begins to discuss with them what it is they have to do, thinking about the whole process even as he speaks.]

MR. TATE: So it's 2 decimal places to the left, not to the front, you were right. I wasn't understanding the question before. This still has to do with growth factor.

[*They concluded that in r = 70 * 0.95x the factor is not 95%, but 5% (subtracting 100-95).* In a similar manner, they talked about 2x as a growth factor; they multiplied it by 100 to turn it into a percent and saw that this was really 200%, subtracting 100%, they get a growth factor of 100% or 1. Once they understand this, Lori and Alberto continue working.]

The vignette provides an example of problem solving that is very different from the standard teaching-learning interaction that occurs in many classrooms. The teacher and students are working collaboratively on what for them is a puzzling problem. They use problem-solving strategies so that they can understand the concepts undergirding the problem, and the teacher and tutor both seem comfortable with expressing that they are not necessarily sure how to proceed. Further, the tutee, Alberto, seems to feel comfortable with their uncertainty. Yet I would argue that all the participants understand better what is going on with exponential growth and decay because they have struggled with these concepts together.

However, on occasion, it was difficult for the teacher to relinquish his standard position as authority and arbiter of mathematics knowledge. In the vignette below, the teacher is called to explain a definition to a tutor–tutee dyad:

Rachel (the tutor) is confused by the directions. She is unsure what *locus* means. We (the advisor and the tutor) call over Mr. Tate to explain the questions. Mr. Tate told Christopher (the tutee) that it is 6 points to the left, 6 points to the right, 6 points up, 6 points, down and also diagonally. So it is a circle. Mr. Tate started on the next problem, which is to provide a description of all points 3 units from (0,-2).

I interrupted him and said that Rachel now understands and can explain (I saw Rachel trying to jump in when Mr. Tate started explaining the circle). (Field notes, March 8)

In this example, the advisor acts on behalf of the tutor when Mr. Tate continues to explain the concept, when the tutor seems ready to continue working with the student.

In another instance, when Coco, Mr. Tate, and a tutee are trying to identify parts of a triangle, Coco admonishes that the tutee should understand the mathematical terms in her own way:

Coco [*talking to tutee*]: Now remember what bi is. Bi is two. So this doesn't divide into two equal parts. So it's not perpendicular bisector but you thought perpendicular since 90 degrees so that's good . . .
Mr. Tate: Yeah, it is an altitude. Altitude gives you 90 degrees.
Coco: She should put in her own words and not just copy.

Some mentors talked explicitly about how their thinking about control and authority relationships between teachers and students had changed as a result of their participating in the program:

> I learned that the students can really like take charge. They don't always need an adult to be teaching them. . . . You can give them some responsibility and they will succeed. (Maria)

> Kids have always collaborated when I was teaching before. But having that whole peer tutoring thing is important, where kids can actually help each other. Being a teacher, sometimes you want to control everything that's in your universe. It's your classroom and you want to have a say about everything that goes on. And sometimes when you—I know, I'm guilty of this all the time—when I see two kids talking in the back of the class, I assume it's about this weekend or what Bobby did yesterday afternoon. But sometimes it is just, he's trying to say, "Let me help you." And that's something I need to really try to foster, and let go, because it is only helping people. (Abby)

Several mentors noted that seeing how tutors explained problems or concepts to tutees was illustrative. On the open-ended questionnaire item asking what would be most helpful to mentors in their future high school teaching Lorraine wrote:

> The methods used by the tutors are most likely the methods that are understood better by the students and hopefully, I, in turn, can use them in the classroom.

In her interview, Lorraine further noted:

> You don't always have to teach students from a lecturing perspective, you kind of get down to their level and work with them through problems and encourage them to use previous knowledge, problem-solving skills.

It was clear that some mentors developed strong teaching-learning relationships with tutors and tutees over time. Initially, at the start of the program, the mentors focused on helping students solve homework problems without engaging them in other aspects of mathematics. This remained the case for some mentors throughout the program (namely, Jane, Candace, Lorraine, and Vera), even when opportunities presented themselves for them to engage with the students on a deeper level. However, Abby, Michelle, and Maria, for example, sought to more deeply engage with students beyond the mathematics students were doing in class. Abby found this difficult initially:

> Well, when I first worked with him [John, a tutee], it was just down to business. This is the problem, I've got a question, let's do this. It was here's a problem, help me get to the answer, basically. You know, it was my first few days there, and maybe I wasn't in the . . . I was trying to get to know the students, maybe I wasn't in the mood to, not *in the mood*. I didn't feel comfortable enough trying to put new math thoughts. You know, they had a question, I had an answer. Maybe I was just trying to form a relationship. But as the semester went on, I felt that the tutees and the tutors and the mentors all started to form a relationship with each other that we felt, the tutees felt comfortable asking questions to anybody and they were also more comfortable taking answers in not so direct form. Sometimes, when kids come to tutoring they just have a question, "I want the answer. Don't try to teach me anything new, or tell me what I did wrong, let's move on." Sometimes that's not the best way to, sometimes that's just filling the bucket of knowledge instead of lighting the fire of knowledge.

As the semester progressed, Abby found it easier to work with students and "light the fire of knowledge":

> I guess I'm saying I was still in the relationship-forming stage and trying to see how far, maybe, I can push them. So, . . . especially with John, he seemed like a very receptive young man. He was very quiet at the beginning too. He wasn't much of a chitchatter. I'm not saying he was a motor mouth at the end but he seemed to come out of his shell more and more as the semester went on, especially with me. . . . He seemed to become more lively as the semester [went on], in that he would seek out help and he would, he was almost excited to come to the sessions, which was great.

Another advisor also commented on the importance of developing relationships with students:

> After working with the same kids, even after just a couple of weeks, they got to know you, they got to know that they could trust you, they could work with you, and I think they almost became more confident in their own ability.

Michelle's conversations with students centered on school-related topics:

I talked to them all the time. I asked them about their classwork, I asked them about their quizzes and tests. You know, some would offer more information than others. But I would ask them you know what or how did the tutoring session help them. And they just said they got their homework done. I imagine that if they went home that they probably wouldn't get their homework done. So, most of the interactions dealt just with school, never anything else. But, most of the time it was just quizzes and tests and how are these sessions helping you.

On a number of occasions, tutees became tutors themselves:

I was working with this student and I think we had gotten to a point where there were 20 exercises and we had kind of worked on most of them. Now it was up to him to finish, like, there was nothing new I could have taught him. Like all of the questions are like the same format, blah blah. What I saw that was interesting, was that the student, a tutee, went to another tutee and was able to help him. I'm not saying he aced his last math test or anything. But for that moment, for that afternoon, for that section, he had understood what he and I talked about so much so that he can go now to another classmate and help him.

CREATING A COLLABORATIVE SPACE FOR DOING MATHEMATICS

The collaborative space that was created by tutors, tutees, advisors, and the teacher over time resulted in productive interactions among high school students. The field notes reveal that the room used for the tutoring site not only served as a place for struggling students to get help, but also as a space for teachers to learn from high school students; for successful students to be exposed to additional mathematics; for students to work on homework by themselves; and, on occasion, for students entering the space to become mathematics experts, regardless of their formal status vis-à-vis the tutoring program. For example, on a day when no tutors came, a tutee who frequently attended sessions, Enrique, served as a tutor for other 9th-graders because he had just taken a test that comprised similar problems to the ones the other students were working on. The advisor writes, "At this point Enrique is confident in his work because Mr. Tate just told him he got a good grade on the test he just took."

At some tutoring sessions, there were distractions in the form of other students who were not participating in the tutoring sessions interrupting. In most cases, this happened because they were waiting for their friends to finish, who were either tutors or tutees. Occasionally, however, these students entered the room and either tutored or received tutoring themselves.

The tutors' confidence in their mathematics and tutoring abilities by the end of the program was apparent—for example, when a student entering the room asked a tutee who to go to for tutoring, Lori immediately interjected, "I'm the best one." One of the advisors, commenting about Rachel's growth as a tutor, wrote, "Rachel is so confident. I am shocked. When I first met her she seemed so shy. Today she walked around the room like a teacher." Students who were not formally participating in the tutoring program valued their interactions with tutors, occasionally commenting on the skills of their tutors to other students. An advisor described such a day:

Today there is another girl in the room who tells Angel that she is a friend of Lori [a peer tutor] and that since they hang out together, her grades have gone up to 90. She encourages him to continue working with Lori.

In addition, advisors positively commented on both tutors' and tutees' increasing confidence with mathematics:

I heard earlier from some girls that, "Ahhh, this is so hard, I can't do this," blah blah blah. And those sorts of statements disappeared as the semester went on. So maybe it could have been hard, I'm sure it was hard. But maybe that "I can't do that" attitude started to disappear. . . . Sometimes students have a complete defeatist attitude with math and, like, there's no way I'm ever going to solve this so why even try. I thought that was something they started to let go of.

Although the collaborative eventually began to show elements of integration of and collaboration among tutors, tutees, advisors, and the teacher, a lingering issue did potentially impede the effectiveness of advisors' work with tutors and tutees: their perceptions of urban schools and the students who attend them.

PRESERVICE TEACHERS' KNOWLEDGE AND BELIEFS ABOUT URBAN HIGH SCHOOLS AND STUDENTS

The mentors had varied perceptions of urban high schools and students. In describing the students at Lowell, Lorraine noted:

I hadn't had urban [school] experiences before, but they were similar to where I was [working before] in central Texas where it was a poor slightly Hispanic community . . . and it was various similarities where—they were not as—education was not their priority.

Lorraine's experiences with students of color (whether they attend rural or urban schools, whether they are "Hispanic" or African American) suggest to her that they have similar beliefs about education. Her hesitation may suggest that it was difficult for her to articulate what she meant by "similar" and that she may be unaware of the different issues affecting urban and rural students of color.

Another mentor, Maria, expressed a seemingly "color-blind" philosophy (Groulx, 2001; Valli, 1995) but did attribute differences between urban and suburban students to their home environments:

> I'm one of those people that believes that no matter where you go, you see the same things. The kids all act the same way. Things that I noticed that were different, it's just the type of homes that some of these children were coming from. It's a little bit more difficult. . . . They didn't have the parental guidance that some students from suburban schools might have.

Elsewhere, Maria repeats her contention about urban students' parents and how teachers are needed to counteract a perceived lack of parental guidance, although she notes that there are some differences in resources available in urban schools:

> It could be more rewarding to work in a school like an urban school because like I said, a lot of these children, they don't really have that parental guidance that—you know. They don't have the same amount of supplies and technology that other schools may have. But I would just tell people that it's more rewarding to work with these children. And they really look up to us more. And they need us so. So we need to be there for them.

Maria's response is an example of what Groulx (2001) refers to as a "missionary" attitude, in that some teachers expect that they will "be helping disadvantaged children through teaching" (p. 79), and that they are needed to address perceived urban students' parental deficiencies.

Despite evidence to the contrary, three advisors expressed overwhelmingly negative perceptions of students' motivation during interviews conducted at the end of the semester. Two also expressed, without any evidence at all, that there was a lack of parental interest and investment in Lowell students' education. At least one advisor, Vera, first mentioned in Chapter 1, continued to express very negative portrayals of urban students despite reporting evidence about positive examples of motivation and diligence on the part of Lowell tutors and tutees. It may be that more time and more experiences are needed so that negative perceptions and stereotypes are critically examined instead of solidified.

Given some advisors' comments during the postinterviews, it is clear that despite their experiences with motivated tutors and tutees, they still equate students' demographic backgrounds as being predictive of students' engagement in school. Without thinking critically about teaching and learning in urban settings and undertaking deeper analysis of the curricular opportunities provided to urban students, it appears that some education students may still be locked into patterns of simplistic thinking about the motivations and interests of urban students. Several advisors made the observation that the mathematics work that students were given at Lowell was not as challenging as it could have been. Jane's and others' observations that "not much is expected or required" of urban students should be a signal to teachers and teacher education students to heighten their expectations of, and requirements of, urban students.

Providing a space for the kind of collaboration that occurred during the Lowell tutoring collaborative was beneficial to those who participated in the effort and suggests that such collaboratives can be positive additions to schools and the learning experience of teachers and students seeking to become teachers. Commenting on the learning interactions that took place, John noted:

> It was great to see how responsive they were and how eager they were to learn. . . . These are kids, and they need to be taught just like all the other kids in the world. So it was a good thing. I saw people there that really wanted to work hard and [were] really trying to better themselves, even though they're climbing a steeper hill than other kids in other districts in other states. They have a much tougher climb than others. And they're working hard at it.
>
> The people who were there were really receptive and really open to learning new things. And even if I hadn't taught them something different they were excited about it. So I really enjoyed it. . . . I learned a lot just from going and learned how well they could work with each other. And that, I guess, I was most surprised about.

Given the problem of African American and Latino/a student underachievement in mathematics, efforts to improve performance should include augmenting the opportunities that students have to learn and do mathematics within mathematics classrooms and, also, in out-of-classroom settings. Although high-achieving high school students provided most of the assistance to their lower-achieving peers, this chapter shows how teachers, graduate students, and high school students can collaborate to develop a community of learners focused on improving mathematics outcomes. The tutoring collaborative described here shows that adolescents can become teachers as well as learners of mathematics, that adults can learn effective instructional strategies from adolescents, and that adults and adolescents can work together to help struggling students outside of traditional authority relationships. In short, a shift in traditional classroom hierarchy in secondary schools need not be accompanied by a loss of respect for the teacher, but rather such a shift can underscore that the ideas, knowledge, and contributions of all the classroom participants are valued. The findings from this study suggest that classroom teachers can enhance their practice by, for example, listening to student discourse to gain alternate methods of explaining concepts and allowing students to lead classroom discussions about mathematics problems, solutions, and ideas. In keeping with reform efforts seeking to shift mathematics teaching and learning from a didactic enterprise centered on the teacher to a dynamic interchange between teachers and students (NCTM, 2000), this project demonstrates that collaborative participants modeled ways of interacting that could be illuminating for teacher educators, administrators, teachers, and high school students interested in improving mathematics outcomes. Most important, it shows that expertise in doing, learning, and teaching mathematics is not limited to adults who excel in the subject, but that these abilities potentially can be developed in demonstrably powerful ways by high school students.

CODA

In the summer of 2006 the principal of Lowell High School, who had been an instrumental supporter of the research conducted at the school and the development of the peer tutoring program, was promoted and left the school for a new position. That school year (2006–2007) the peer tutoring program was not implemented. Three years and two principals later, I approached the current teaching staff and Lowell's administration about reinstituting the peer tutoring program. The current teaching staff, which included only one teacher who had been at the school in the 2005–2006 school year, and the new principal and assistant principal were interested but noncommittal. A continuing assistant principal, who had also been at the school in 2005–2006, championed the project and encouraged the teachers and new administration to institute it.

Teachers recruited tutors, and the first training session I conducted in November 2009 had 20 students, a dramatic increase from the 4 tutors I had trained several years earlier. Also, unlike the first incarnation of the tutoring program, the 20 tutors were predominantly male. I mentioned to the administrative staff and teachers that in the 2005–2006 incarnation, the program had taken a few weeks to grow but by midsemester there were many students participating. The decision was made to offer the program two days a week after school.

Unlike the previous program, the 2009–2010 version of the peer tutoring program took off immediately. The tutoring classroom was filled with students, tutors, and tutees. Students at the school requested that the program run 3 days per week instead of 2. Another classroom had to be used to accommodate the overflow of students. Instead of one supervising teacher, 3 of the 4 math teachers at the school became involved in the program. I was asked to come train a new batch of tutors.

The new assistant principal, who made the request, said, "I'm surprised. I didn't expect this many students to be interested in something like this."

While the program undeniably benefited from a particularly charismatic teacher who drew a number of students after school, the students' enthusiasm and interest in the program was undeniable. Students' grades improved, and the math classroom became *the* place to be after school, at least for part of the afternoon.

But innovative ideas by the participants (students and teachers) were stymied. When students asked that a during-the-school-day component be launched during the lunch hour, the administrative staff wondered where such tutoring would take place. Space was at a premium: The library was being occupied by students during lunch and the lunchroom was deemed too noisy. Some teachers offered their classrooms for tutoring during lunch. But the question of supervision arose: Which teacher or staff member would supervise the students during the day? How would they be compensated for this extra duty?

These kinds of issues hindered the growth of an exciting development at the school: Students were interested in math, and were interested in doing more of it. The opportunity to shift the peer tutoring program from a focus on homework and test preparation to emphasize enrichment and learning of new mathematical ideas was at hand, but frustratingly out of reach because of logistical and organizational concerns.

As others have pointed out, the sustainability of promising and effective programs, and their dependence often on supportive administrators and dedicated individuals working with these programs, remain an issue. Despite its promise for improving students' mathematics outcomes (including performance and socialization), and the high engagement levels of students and teachers, the peer tutoring program has operated formally at Lowell only in 3 years out of the last 6. Reports from teachers and students indicate that students continue the work of peer tutoring even when a formal program is not in place, but the goal of establishing a formal program was to institutionalize positive practices of students around mathematics, to expose students and teachers to the power of peer communities working together in mathematics, and to ensure that a dedicated space at the school existed for the support of students' mathematics work. While it is exciting that elements of the program have continued to exist "underground," ensuring that effective programs are supported and sustained by administrators, teachers, and students is a critical part of improving mathematics achievement in the long run.

CHAPTER 6

Conclusions

A QUESTION OF OPPORTUNITY

The struggle to sustain, continue, and expand the work of programs and initiatives that support underserved students' learning in mathematics raises questions about our commitment to creating lasting and durable environments that provide high-quality mathematics education for all. In particular, as educators we must consider the following questions: What opportunities are present in urban schools to build students' interest and excitement about math? What opportunities are present in urban schools to facilitate communities of learning around math? And how often are these opportunities ignored? What implications do these opportunities, and the ignoring of them, have for students' success?

This book has been an answer to these questions. Throughout, I have emphasized that urban students are interested in learning and in mathematics, and that they come from communities and networks that are committed to their education and mathematics development that have gone unnoticed and unacknowledged. I have placed at the forefront Black and Latino/a students' voices and experiences, because these are so often missing from discussions about teaching, learning, and achievement. I have described the critical roles of schools, teachers, and programs that build on students' strengths, rather than focus on students' weaknesses.

When we reconceptualize mathematics teaching and learning in schools in these ways, it requires that we consider new and far-ranging implications for policy and practice. In particular, I wish to focus on the implications of this work for teachers, teacher educators, administrators, and policy makers.

IMPLICATIONS FOR TEACHERS

Teachers should be aware that mathematics talent and interest come in many forms. Narrow interpretations of the talented mathematics student—quiet, a quick completer of exercises, a high performer on multiple-choice examinations—may restrict teachers' identification of students with potential to do well in mathematics. For example, Lowell High students readily identified several peers whom they deemed high achieving—and indeed, were high achieving upon confirmation of their grades and test scores—who were not initially identified as such by their teachers. Indeed, students were able to point to specific mathematics behaviors beyond test scores that are measures of mathematics competence. Further, mathematics potential in all students is worth developing, and teachers' assumptions and attendant behaviors can either stymie or promote that development.

Teachers also need to be aware that their instructional behaviors signal their interest in students' mathematics learning, and that students interpret these behaviors as indicators of teachers' perceptions and expectations (or lack thereof) for student success. Students speak very compellingly, honestly, and knowledgeably about the work of teachers. They are not expecting friendships but do expect (and respond well) to teachers who exhibit caring about and interest in them and their learning and success. When teachers hold low expectations of students, high school students in particular are aware of these. The research indicates (but not necessarily for teachers at Lowell) that generally teachers may hold limiting views of students based on their racial/ethnic and social class backgrounds. Teachers must be aware that their practices with students (for example, assigning low-level mathematics work repeatedly, not exposing students to engaging problems) may reflect their stereotypical notions of students' interest, competence, and potential, and that these practices can impede student progress and have cumulative and long-ranging impact on students' life outcomes.

Further, students report that most of their mathematics classroom experiences have been unengaging. They deem engaging experiences to include instruction that incorporates the linking of mathematics content to meaningful real-world contexts, for example, as well as exposure to challenging mathematics, rather than instruction that is solely focused on learning of procedures and note-taking. Paradoxically, many students report that they like a great deal of explication of content and may initially balk when teachers provide opportunities for exploration and discovery; this may be due to students' conceptions of mathematics as a series of procedures. However, it is important to note that this perspective has been reinforced by most of their mathematics experiences in school. Teachers should also be aware of the social aspects of learning, in particular, how students' peer groups support and do not support mathematics learning—and how some of these differences are rooted in students' cultural and gender backgrounds. They should be aware that students' peer group support for academic and mathematics behaviors is complex and nuanced—and that lower-achieving peers can be important supporters of success. More broadly, teachers should be aware that there are extensive support networks for students' mathematics learning that incorporate extended family members, peers, adults with little formal education beyond high school, and previous teachers.

The peer tutoring program I described in the previous chapter and other initiatives that incorporate peer academic support build on this notion of community support for learning. While maintaining teachers' authority in the classroom is important, this work as well as that of others supports the notion of repositioning students in the classroom as worthy contributors to the development of mathematics knowledge and understanding. In particular, promoting discussion and group problem solving in the classroom increases students' agency and active interest in their own mathematics learning, and ensures the development of metacognitive and problem-solving skills on which students can rely when they are participating in lifelong mathematics endeavors.

IMPLICATIONS FOR TEACHER EDUCATORS

Teacher educators have to effectively address the troubling stereotypes that novice and experienced teachers hold about urban students and their intellect, potential, and interests. This must be done carefully and thoughtfully, to avoid reifying stereotypical notions of who can do mathematics. Teacher educators should ensure that prospective teachers are exposed to student teaching and clinical experiences with urban students that are not all focused on remediation and "struggle," but rather experiences with urban students that reveal those students' competence, excellence, and interest in mathematics. Teacher educators also must be purposeful in their discussions about gaps in demonstrated performance in mathematics (the "achievement gap") and have attendant discussions about the real disparities in opportunities and resources that contribute to it.

Teacher educators who are preparing secondary mathematics teachers also have to address what appears to be a critical gap between teachers' beliefs and practices around mathematics. Mathematics teachers report beliefs that appear to endorse mathematics as a discipline of interconnected topics that requires thinking and understanding about mathematical problems, but endorse and engage in practices that reinforce common views of school mathematics as a series of disconnected, discrete topics that require speed and efficiency in completing procedures. While procedural fluency is necessary to mathematics learning, conceptual understanding is also important and it is clear from extensive research in mathematics classrooms that U.S. students are not generally getting much of the latter in secondary mathematics classrooms. Some teachers report a lack of exposure to reform mathematics practices in their preservice teacher education programs, and teachers themselves continue to be taught mathematics in ways that support narrowly defined views of mathematics learning and teaching, both in their experiences as secondary school students themselves as well as in their college and graduate school careers. Ensuring that preservice teachers are supported in learning to teach in ways that support the spirit of mathematics reform and focus on equitable teaching practices, as well as ensuring that students have clinical experiences beyond student teaching that support these goals, are important components of teacher education programs. There is some evidence that teachers find it difficult to engage in these kinds of practices because of their lack of exposure to them in preservice teacher education programs;

concerns about time and classroom management; and concerns about and interpretations of accountability measures, such as standardized tests and teacher evaluations, that appear to reward narrow conceptions of mathematical knowledge and skill, and mathematics teaching that is focused on procedures.

IMPLICATIONS FOR ADMINISTRATORS AND POLICYMAKERS

As discussed throughout Chapter 3, policymakers are not just individuals at the district and state level far removed from the mathematics classroom; teachers and administrators are policymakers, too. For example, at the school level, the decisions that teachers and administrators make about offering (or not offering) advanced mathematics courses have dire consequences for students. Often these expectations are rooted in stereotypes and misunderstandings about what is the most appropriate curriculum for urban students (Haberman, 1991). As the work of Burris et al. (2006), Gutiérrez (1999, 2000), Werkima and Case (2005), Escalante and Dirmann (1990), Moses and Cobb (2001), and others reveal, ensuring access to advanced mathematics for secondary students results in improved outcomes. When done thoughtfully and with rigor and care, students who previously demonstrated low achievement can achieve at high levels. Finally, Useem's (1992) work puts into stark relief that district-level administrators' disparate conceptions of mathematics and what is "appropriate" for certain students can have a critical impact on students' opportunities to learn mathematics. These conceptions, and their attendant decisions, may not be based on anything other than administrators' erroneous perceptions and "gut feelings."

Finally, it is worth stating that improvement in mathematics performance takes time and investment and is not an overnight proposition. This flies in the face of policies that demand improvement in months, rather than years. It is not to suggest that we have unlimited time to make improvements in mathematics education—we don't—but rather that as long as teachers and administrators are transparent about well-conceived plans and policies to improve student achievement, progress (measured in multiple ways) should be taken into account. As discussed in Chapter 1, Jaime Escalante's AP calculus program took 8 years to demonstrate the patterns of success we now all know about.

During the past 30 years, a great deal of attention has been paid to developing standards for mathematics teaching and learning. Crafting high standards for achievement and performance is a worthy goal, and standards are good benchmarks for teachers, students, and parents to have. But as Tate (1997) and others note, without serious attention to resources needed not just to develop standards but also to implement them, outcomes related to those standards will continue to reflect disparities in resources, opportunity, and access. In addition, the tension between local control of education and national priorities has resulted in a great deal of variety in which standards are emphasized and how they are implemented in different schools, districts, and states. The adoption of the Common Core State Standards by more than 40 states might suggest progress in identifying a unified set of standards for localities; however, it should be emphasized that standards generally are *minimum* benchmarks for proficiency in a subject area. In some ways, we hope and expect students to not only meet but exceed standards. Despite states' adoption of the Common Core State Standards, there will still be differing understandings across schools and districts about which ones should be emphasized and deemphasized. There will, without attention to implementation, still be a wide range of interpretations of resources needed to help students meet and exceed standards (and teachers to teach to those standards), especially in urban schools.

In addition, the adoption of standards has the potential to contribute to a redefinition of what mathematics is and for what purposes mathematics education in school occurs. This was certainly the case with the 1989 NCTM Standards, which spawned numerous discussions and debates about "conventional" versus "reform" mathematics curricula and teaching. In addition, new kinds of assessments that incorporated opportunities for students to demonstrate their mathematics knowledge beyond static multiple-choice responses emerged. However, assessment, instruction, and curriculum influence one another, and the implementation of NCLB in some ways has had a narrowing effect on how we think about mathematics learning. Districts have reverted to assessments that primarily measure basic skills and elementary concepts. The Common Core State Standards are an attempt to redefine standards for mathematics education once again. In aiming for "clarity and specificity," it is possible that the attendant curricula and assessments spawned by the Common Core State Standards may have a reductive impact on how we think about mathematics in school.

This would be unfortunate, because the voices of students (both current high school students and mathematicians reflecting on their adolescent experiences) reveal that some of the most pivotal moments for their mathematics interest and learning occurred when they were exposed to mathematics that was unique and different from the typical textbook or worksheet exercise. Some of these moments were quite small in scope and derived from a brief conversation with a teacher outside of class about a mathematics concept, or a novel problem that incorporated mathematics content already "covered." The work of the peer tutors in the Lowell peer tutoring collaborative revealed that students draw upon pedagogical strategies and interconnected content to help their struggling peers. To their peers, these explanations were novel and provided a new way of thinking about mathematics problems, even those problems that were not particularly complex. In addition, both high school students and mathematicians describe "out-of-school" experiences that are formative for their mathematics interest and knowledge. Many elements of these out-of-school experiences could, and should, be incorporated in some way in school mathematics courses. But if administrators and policymakers see mathematics in limiting ways, teachers will be unable to make the case for incorporating some of these meaningful activities in classrooms.

IMPLICATIONS FOR RECONCEPTUALIZING
STUDENT ACHIEVEMENT IN MATHEMATICS

This brings us to reconceptualizing how we think about student achievement in mathematics. Our societal views of mathematics as a difficult subject that very few people do well in, I have argued, influence how we conceptualize appropriate teaching, learning, and assessment. These narrow definitions of what it means to be good in mathematics—and what it means to be a good mathematics student—permeate teachers' conceptions of students and students' conceptions of themselves. These definitions are underscored, perhaps inadvertently, by test developers and a current accountability climate that rewards basic skills and procedural fluency while paying little attention to students' problem-solving abilities and critical thinking skills. In a circular fashion, teachers begin to adapt their teaching practices to support success on tests that often measure disconnected and esoteric skills that require little thinking and do not fully assess students' mathematical knowledge. Too often, students are being taught mathematics that is aligned with the basic skills requirements of the exams, not the advanced skills required for college and the problem-solving requirements of real life.

Arguably no one test can measure all the important aspects of mathematical knowledge that we deem important. But it has to be reiterated that mathematical knowledge is not limited to what students can demonstrate on an examination. And learning is affected by important affective measures that are often unmentioned in discussions about student learning, performance, and achievement. From much of the research, it appears that the sole purpose of testing is to document who is failing, and in relationship to whom. By painting students who do not perform well on these assessments as "low achieving" and not ensuring that they are exposed to opportunities in school to engage in mathematics critically and meaningfully, we separate the original intent of testing as a means to evaluate where students are—and do something about it—from these results. Too much of the research literature and policy statements around mathematics performance portray urban students, and in particular Black and Latino/a students, in narrow ways that do not account for important differences within these groups. Research that explores differences within and across racial/ethnic groups makes clear that there are critical differences within and across groups in mathematics behaviors, performance, persistence, and achievement. Gender differences in attitudes toward mathematics, for example, appear to be larger for Latino/a students than for Black students. In addition, our current use of testing as a broad yardstick measure of student and school performance ensures that we spend an inordinate amount of time discussing failure, rather than deeply examining causes of student success and how students' networks may support and obstruct it.

Throughout this book I have suggested that schools have a great deal to learn from the experiences of those who have been successful in mathematics through adulthood (particularly the mathematicians I have mentioned) as well as successful college programs in science, technology, engineering, and mathematics (STEM) that target underrepresented students. The best of these programs incorporate several key concepts that bear repeating: First, they assume students are excellent and competent—they are not remedial programs (Steele, 1997; Treisman, 1992). Second, they incorporate academically supportive peer groups that foster learning and engagement. Third, they are sustainable—most often sustained by committed leadership, but also by participants who carry out the mission of the program within the program's confines and beyond them. These programs and experiences are both formal and informal and occupy spaces that cross important boundaries—they encompass activities within and outside of classrooms, within formal educational institutions and within neighborhoods, among novices and experts, and among peers and mentors.

The development of more mathematical spaces like these that are available for Black and Latino/a youth in urban schools is critical to ensure that we are developing not just mathematical competence, but also mathematical excellence, across all populations. I identify mathematical spaces as sites where mathematics knowledge is developed, as sites where induction into a particular community of mathematics doers occurs, and as sites where relationships or interactions contribute to the development of a mathematics identity. These spaces, then, may be physical locations like a school or classroom, or they may be locations to which the individual attaches a particular social, cultural, or mathematical meaning because of interactions and experiences he or she has there (Walker, 2010).

In our work with urban students, I suggest that we move from solely sometimes "inadvertent" spaces that foster development for individuals (like individual positive teacher-student interactions and informal peer-to-peer discussions about mathematics) to creating and examining "intentional" spaces both within and outside of classrooms, like peer tutoring collaboratives, math circles, and other programs that contribute in strong ways to mathematics socialization and talent development for larger groups, particularly for underserved students. Efforts to craft purposeful mathematical spaces for Latino/a and African American students, I argue, should challenge the pervasive discourses of deficiency that permeate discussions of Black and Latino/a mathematics achievement (Stinson, 2006; Martin, 2009b) and reflect the bridging of out-of-school and in-school networks, relationships, and practices that exist within these communities. A recent example profiled in the *New York Times*, the Bard College Summer Program in Mathematical Problem Solving, was designed to provide low-income New Yorkers gifted in mathematics a 3-week sleepaway camp experience centered on mathematics. As the director of the program noted, the point is not to "offer remedial instruction to struggling students, but rather to challenge those who already excel" (Cromidas, 2011). In addition, the camp provides an important socialization experience: it is an opportunity for students to meet other young people like them, who are interested in mathematics, and perhaps an initial comfortable foray into the world of math circles and math competitions, early experiences that are important to the development of the mathematically talented.

We can be much more successful in improving mathematics outcomes and fostering interest and engagement in mathematics for urban students. But this does require that we rethink how and where mathematics teaching and learning occur, what denotes talent and interest in mathematics, and who can be excellent in mathematics. To do this, we should build on existing out-of-school spaces that support mathematics socialization and also reimagine the mathematics classroom to be a space that supports mathematics identity development and positive socialization experiences as well as one that provides opportunities to learn meaningful mathematics.

But we also have to think about how we ensure that meaningful mathematics occurs beyond fleeting conversations, students' individual experiences and identities, and the spaces in which they happen to find themselves. Our expectations of students' abilities are key—if we think students have potential and if they are worthy of our attention in spaces that support mathematics learning, we become much more intentional and purposeful about creating these spaces. Thus, opportunities to engage in meaningful mathematics have to have intentionality and purpose and should not solely be haphazard or happenstance. For too many of our students, particularly our underserved Black and Latino/a students, these opportunities are limited. It is my hope that this book provides some answers to the questions of how to build on students' existing strengths and interests in mathematics, craft intentional and meaningful communities and spaces that promote mathematics engagement and socialization, enhance mathematics teaching and learning in schools, and thus ensure meaningful mathematics learning for all.

Methodological Notes

A significant portion of the analysis presented in *Building Mathematics Learning Communities* is garnered from a two-phase mixed methods research project I conducted at Lowell High School between 2004 and 2007. Using qualitative and quantitative research methodologies, I explored questions about students' mathematics attitudes, participation, and performance as well as the impact that students' networks (peers, families, teachers, and others) had on those outcomes. In 2004–2005, Lowell High School (Lowell High School and all names of students are pseudonyms) was a small public high school in New York City, serving approximately 300 students in grades 9 through 12. About 97% of the students attending Lowell were Latino/a (56%) or Black (41%). The student body was predominantly female (60%), with about 70% of the students qualifying for free or reduced-price lunch. Most of the students attending Lowell came from Upper Manhattan, which is one of the least economically advantaged but most culturally rich regions of the city of New York. The school was highly valued among neighborhood parents for its small size and dedicated corps of teachers. About 80% of Lowell's 20 teachers were fully certified; many (about 74%) had advanced degrees. About half the teachers had spent more than 2 years teaching at Lowell, but most had taught fewer than 5 years in total.

Lowell's average performance on the New York State Regents examination, a required battery of assessments that New York students must take to graduate from high school, was similar to that of many other high schools in the city with similar demographic composition. In mathematics, 65% of Lowell's students scored above a 55 (the range is 0 to 100) on the Regents examination, thus meeting basic graduation requirements; however, only 26% of the students met the requirements to earn the more prestigious Regents-endorsed diploma in 2003, instead of the alternative awarded to students who score above a 55, the local high school diploma. Thus while Lowell's administrators considered it to be a "safe and thoughtful community of learners," with the goal of "meet[ing] the academic and affective needs of students through a rigorous and engaging interdisciplinary Regents-based curriculum" (taken from the Lowell High School Annual Report, 2002–2003), it was the desire of the administrators and teachers to improve students' academic achievement, particularly in mathematics.

The first phase of the project was a study designed to determine teachers' beliefs and expectations about students' mathematical ability and potential, as well as students' beliefs and networks that supported their mathematics engagement and achievement.

During this phase of the project, I conducted (1) a survey of teachers eliciting their beliefs about their students' academic and mathematics potential; (2) a survey of students eliciting their beliefs about and practices around mathematics, and the perceived influence of peers on those beliefs and practices; and (3) interviews of high-achieving students in mathematics, focusing on what they believed to be key influences on their mathematics achievement. Twenty-three Lowell teachers and administrators participated in the teacher survey in fall 2004, and 154 students completed the student questionnaire in spring 2004. The students were predominantly female (58.4%) and Latino/a (55.6%), with 29.4% of students identifying themselves as African American, 9.7% as multiracial, and 2% as other. (Percentages do not add to 100% because 9 students did not report their gender and 15 did not report their ethnicity).

For the student interview component of the study, I asked Lowell's mathematics teachers to nominate their highest-achieving students for participation. Sixteen students were nominated by their teachers. Because teachers may be more likely to nominate students whom they perceive as well behaved and least likely to cause trouble rather than or in addition to high achieving, students participating during the first round of interviews were asked to identify other students who did well in mathematics. Their nominees' achievement was confirmed with the Lowell teachers. This combination of nomination and snowball sampling (Lincoln & Guba, 1985) yielded a final group of 21 students. Several rounds of interviews were conducted in spring 2004, summer 2004, and fall 2004. Interview questions included the following:

> Why do you think you are doing well in mathematics this semester? Have you always gotten good grades in mathematics? If so, why? If not, why not?
>
> Think about your experiences in mathematics in school. Have there been obstacles (persons or things getting in your way) at school that could have prevented you doing well in math? If so, what were they and how did you get around them?
>
> Who or what contributes to your success in math?

Teachers were allowed to nominate students based on their own definitions of "high achieving," but all teachers and students identified students who had earned grades of B or better in their courses. At the time of the interviews, all the students were in grades 9 through 12, ranging in age from 14 to 18.

The sample of students was, as in Lowell High, predominantly female (14 of the 21 participants were female). Eleven of the students identified themselves as Latino or Latina (encompassing Dominican [2], Puerto Rican [5], and Guatemalan

[1] heritage), eight as Black (one of West Indian heritage), and two as African American and Latino/a.

Once the student sample was identified, students participated in an hour-long semistructured interview, which a graduate student researcher conducted using a protocol. The interviews were audiotaped. Although the protocol included detailed questions and cues to the interviewer, interviews often included issues brought up by participants that were not included on the protocol. As part of the interview, students completed a "map" of influences they deemed pertinent to their mathematics success (see Appendix B for an example). I designed this map to elicit who students felt, in addition to themselves, were most responsible for their mathematics success. Using the map in addition to the interview questions helped to elicit additional supporters whom students may not have mentioned in their interview. In addition, students provided additional information about topics mentioned during the interview. Students were asked to provide descriptive information about influences listed.

The second phase of this project used the findings about high-achieving students' peer networks and teachers' expectations to design a peer tutoring collaborative and to examine its impact on preservice and in-service teachers' pedagogical beliefs, knowledge, and expectations as well as students' mathematical knowledge, learning, and behaviors. Thus, the research project data in this book come largely from student achievement data; analyses of questionnaires of teachers and students; in-depth, semistructured interviews (audiotaped) conducted with high-achieving students and preservice teachers; and extensive field notes and observations of classrooms and the peer tutoring project.

Sample Student Map of Influences

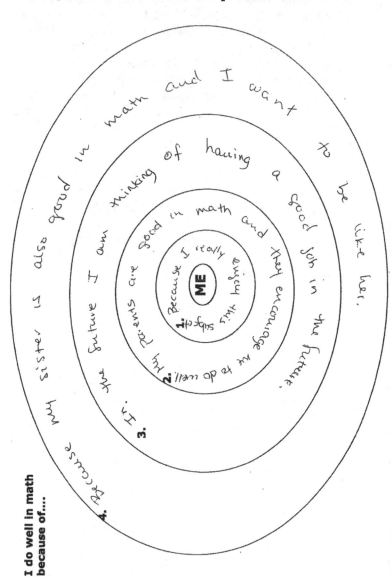

I do well in math because of....

Please label each circle with the reasons you think you do well in mathematics (the circle closest to you would be the reason you think is most important). These reasons can include people, activities, clubs, etc. Include in the circle a short description of how each reason is related to your math success. [If you need additional room, please write on the reverse].

Student-Designed
Recruitment Materials

*don't like math?
*struggling with math?
*ready for the regents?

We know how it feels!
Come get help from your classmates
who are doing great in math now!

Math sessions for 9th and 10th graders
start this Thursday, and continue on
Tuesdays, Wednesdays, and Thursdays
this semester from 3-5.

*Sign up in Mr. XXX's room
*Ask Shanelle B. or Tiffany P.
  for more information.

---

got math?

Need help with your math homework?
Want to get ready for the Regents?
Come get **MATH HELP** from your
classmates who are doing great in math
now!

Drop in hours:
**Tuesdays, Wednesdays, and
Thursdays anytime between 3-5 PM**

got math?

Need help with your math homework?
Want to get ready for the Regents?
Come get **MATH HELP** from your
classmates who are doing great in math
now!

Drop in hours:
**Tuesdays, Wednesdays, and
Thursdays anytime between 3-5 PM**

got math?

Need help with your math homework?
Want to get ready for the Regents?
Come get **MATH HELP** from your
classmates who are doing great in math
now!

Drop in hours:
**Tuesdays, Wednesdays, and
Thursdays anytime between 3-5 PM**

got math?

Need help with your math homework?
Want to get ready for the Regents?
Come get **MATH HELP** from your
classmates who are doing great in math
now!

Drop in hours:
**Tuesdays, Wednesdays, and
Thursdays anytime between 3-5 PM**

Peer Tutor Training Materials

LOWELL SCHOOL PEER TUTORING ORIENTATION

1 February 2006

I. Introductions
II. Overview
III. Questionnaire
IV. Problem Solving

My sister is extremely forgetful. On her last New York visit, she told me that her new office phone number ended with "1983" or maybe "9183."

"Well, which is it?" I asked.

My sister responded, "Well, I think it's 1-9-8-3, but what does it matter? I know there's a 9, an 8, a 1, and a 3 in the number somewhere. There are only a few other combinations. You can try all of them."

Think about this: Are there really "only a few" combinations possible with those four digits? How many telephone "suffixes" with 1, 9, 8, 3 are possible? Four? More than 10? More than 20? Explain your reasoning.

Solve this: Suppose Dr. Walker knew that the first digit of her sister's suffix was "1." How many possible combinations are there now?

Polya's 4-Step Process for Solving Problems

1. Understand the problem.
2. Devise a plan (can one or more of the strategies below be used?)

Guess and test	Use direct reasoning	Look for a formula
Use a variable	Use indirect reasoning	Do a simulation
Draw a picture	Use properties of numbers	Use a model
Look for a pattern	Solve an equivalent problem	Use dimensional analysis
Make a list	Work backward	Identify subgoals
Solve a simpler problem	Use cases	Use coordinates
Draw a diagram	Solve an equation	Use symmetry

3. Carry out the plan.
4. Look back.

A worker placed white tiles around black tiles in the pattern shown in the three figures below (adapted from the Massachusetts Comprehensive Assessment System-MCAS):

 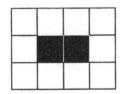 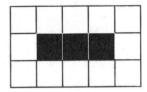

a. Based on this pattern, how many white tiles would be needed for four black tiles?

ANSWER: _____

b. How many white tiles are needed for 14 black tiles? Explain how you got your answer.

ANSWER: _____

Looking at Sample Responses

First, solve the following problems that were on a recent Lowell math exam. Then we'll look at some students' solutions.

1. Solve: $x + 3 = 6$

2. Compare the quantities in Column A and Column B.

Column A	Column B
The solution to $y - 6.5 = 7.8$	The solution to $y + 6.5 = -7.8$

   [a] The quantity in Column A is greater.
   [b] The quantity in Column B is greater.
   [c] The quantities are equal.
   [d] The relationship cannot be determined from the information given.

3. Solve for the variable: $8x + 4 = 36$

4. Solve for the variable: $-8n + 26 + 10n = 54$

5. Multiply: $(x + 2)(-8)$

Tips for Tutoring

1. Be helpful, but don't do the work for your peer. Sometimes showing them how to do a problem is necessary, but the idea is for them to be able to do the work without you.
2. Watch your tone when you're helping someone. They have math knowledge, some of it is incorrect, but some of it is correct, too. Don't shut them down by belittling them or their solution process.
3. Remember the problem-solving strategies, and remind your peer of them too. You can say something like, "This was hard for me too, until I kept practicing it." Encouragement is important!
4. Others?

Notes

Introduction

1. A description of the research conducted at Lowell High School is in Appendix A. The names of the school, teachers, and students are pseudonyms.

Chapter 1

1. For a thoughtful and extensive discussion of the performance gap, see Jencks and Phillips (1998).

2. For discussions of inequitable opportunities to learn mathematics, see, for example, Oakes (1990) and Tate (1995).

3. Additionally, the high national dropout rates for Native American and Latino/a students affect their persistence in mathematics. Nationally, in the 1990s only about 57% of Latino/as and 63% of Native Americans complete high school. White and Black American high school completion rates are comparable, around 87%. Asian American students' completion rates range from 88% (immigrant) to 95% (native-born) (College Board, 1999).

4. See Walker (2001) for an extensive bibliography pertaining to racial-ethnic differences in mathematics participation.

Chapter 2

1. Elements of this chapter originally appeared in the *American Educational Research Journal* (see Walker, 2006).

2. Damon was not interviewed for the study of high-achieving students at Lowell.

Chapter 3

1. These are all comments I have heard in my years as a mathematics teacher and mathematics education researcher.

2. As described in Chapter 1, these statements are largely unsupported by fact and reflect pervasive societal stereotypes.

Chapter 5

1. The peer tutoring program has operated on and off at Lowell since 2006. The bulk of the data presented here come from its inaugural year, 2005–2006, with some achievement data from 2009–2010.

2. The mathematics curriculum in New York State changed from Math A/Math B to separated content areas, including integrated algebra, geometry, and trigonometry, in the 2008–2009 school year. In 2011, the New York State Board of Regents approved a traditional course pathway for high school mathematics courses (including Algebra I, Geometry, and Algebra II), adhering to the Common Core State Standards guidelines.

References

Ainsworth-Darnell, J. W., & Downey, D. B. (1998). Assessing the oppositional culture explanation for racial/ethnic differences in school performance. *American Sociological Review, 63,* 536–553.

Allexsaht-Snider, M., & Hart, L. E. (2001). Mathematics for all: How do we get there? *Theory into Practice, 40*(2), 93–101.

American Association of University Women (AAUW). (1992). *The AAUW report: How schools shortchange girls.* Washington, DC: AAUW Education Foundation and National Education Association.

Anderson, J. (1988). *The education of Blacks in the South, 1860–1935.* Chapel Hill: University of North Carolina Press.

Anyon, J. (1997). *Ghetto schooling: A political economy of urban education reform.* New York: Teachers College Press.

Anyon, J. (2005). *Radical possibilities: Public policy, urban education, and a new social movement.* New York: Routledge.

Appelbaum, P. M. (1995). *Popular culture, educational discourse, and mathematics.* Albany: State University of New York Press.

Auerbach, S. (2002). "Why do they give the good classes to some and not to others?" Latino parent narratives of struggle in a college access program. *Teachers College Record, 104,* 1369–1392.

Azmitia, M., & Cooper, C. R. (2001). Good or bad? Peer influences on Latino and European American adolescents' pathways through school. *Journal of Education for Students Placed at Risk, 6,* 45–71.

Bakari, R. (2003). Pre-service teachers' attitudes toward teaching African American students. *Urban Education, 38*(6), 640–654.

Bell, L. A. (2002). Sincere fictions: The pedagogical challenges of preparing White teachers for multicultural classrooms. *Equity and Excellence in Education, 35*(3), 236–244.

Bempechat, J. (1998). *Against the odds: How "at-risk" children exceed expectations.* San Francisco: Jossey-Bass.

Berry, R. Q. (2008). Access to upper-level mathematics: The stories of successful African American middle school boys. *Journal for Research in Mathematics Education, 39*(5), 464–488.

Boaler, J. (2002). The development of disciplinary relationships: Knowledge, practice, and identity in mathematics classrooms. *For the Learning of Mathematics, 22,* 42–47.

Boaler, J. (2006). Urban success: A multidimensional mathematics approach with equitable outcomes. *Phi Delta Kappan, 87*(5), 364–369.

Boaler, J., & Greeno, J. (2000). Identity, agency, and knowing in mathematical worlds. In J. Boaler (Ed.), *Multiple perspectives on mathematics teaching and learning* (pp. 171–200). Stamford, CT: Ablex.

Boaler, J., & Staples, M. (2008). Creating mathematical futures through an equitable teaching approach: The case of Railside School. *Teachers College Record, 110*, 608–645.

Bol, L., & Berry, R. Q. (2005). Secondary mathematics teachers' perceptions of the achievement gap. *The High School Journal, 88*(4), 32–45.

Breitborde, M. (2002). Lessons learned in an urban school: Preparing teachers for the educational village. *The Teacher Educator, 38*(1), 34–46.

Burris, C. C., Heubert, J., & Levin, H. (2006). Accelerating mathematics achievement using heterogeneous grouping. *American Educational Research Journal, 43*(1), 103–134.

Burton, L. (1999). The practices of mathematicians: What do they tell us about coming to know mathematics? *Educational Studies in Mathematics, 37*, 121–143.

Cahnmann, M. S., & Remillard, J. T. (2002). What counts and how: Mathematics teaching in culturally, linguistically, and socioeconomically diverse urban settings. *The Urban Review, 34*(3), 179–204.

Campbell, P. (1995). Redefining the girl problem in mathematics. In W. Secada & E. Fennema (Eds.), *New directions for equity in mathematics education* (pp.225–241). New York: Cambridge University Press.

Cammarota, J. (2004). The gender and racialized pathways of Latina and Latino youth: Different struggles, different resistances in the urban context. *Anthropology and Education Quarterly, 35*, 53–74.

Castro, A. J. (2010). Themes in the research on preservice teachers' views of cultural diversity since 1985: Implications for researching millennial preservice teachers. *Educational Researcher, 39*(3), 198–210.

Catsambis, S. (1994). The path to math: Gender and racial-ethnic differences in mathematics participation from middle school to high school. *Sociology of Education, 67*, 199–215.

Checkley, K. (2001). Algebra and activism: Removing the shackles of low expectations—a conversation with Robert P. Moses. *Educational Leadership, 59*(2), 6–11.

Cirillo, M., & Herbel-Eisenmann, B. (2011). Mathematicians would say it this way: An investigation of teachers' framings of mathematicians. *School Science and Mathematicians, 111*(2), 68–78.

Cobb, P., & Hodge, L. L. (2002). A relational perspective on issues of cultural diversity and equity as they play out in the mathematics classroom. *Mathematical Thinking and Learning, 4*, 249–284.

Conchas, G. Q. (2001). Structuring failure and success: Understanding the variability in Latino school engagement. *Harvard Educational Review, 71*, 475–504.

Cook, P. J., & Ludwig, J. (1998). The burden of "acting White": Do Black adolescents disparage academic achievement? In C. Jencks & M. Phillips (Eds.), *The Black–White test score gap* (pp. 375–400). Washington, DC: Brookings Institution.

Cooper, C. W. (2003). The detrimental impact of teacher bias: Lessons learned from the standpoint of African American mothers. *Teacher Education Quarterly, 30*(2), 102–116.

Cooper, C. R., Cooper, R. G., Azmitia, M., Chavira, G., & Gullatt, Y. (2002). Bridging multiple worlds: How African American and Latino youth in academic outreach programs navigate math pathways to college. *Applied Developmental Science, 6*, 73–87.

Corbett, D., & Wilson, B. (2002). What urban students say about good teaching. *Educational Leadership, 60*(1), 18–22.

Corbin, K. & Pruitt, R. (1997). Who am I? The development of the African American male identity. In Vernon Polite (Ed.), *African American males in school and society: Practices and policies for effective education* (pp. 68–81). New York: Teacher College Press.

Cromidas, R. (2011, July 27). A sleepaway camp where math is the main sport. *New York Times.* Retrieved from http://www.nytimes.com/2011/07/28/nyregion/a-sleepaway-camp-for-low-income-ny-math-whizzes.html

Crosnoe, R. (2006). *Mexican roots, American schools.* Palo Alto, CA: Stanford University Press.

Cushman, K. (2009). SAT Bronx: A collaborative inquiry into the insider knowledge of urban youth. *Theory into Practice, 48*(3), 184–190.

Darling-Hammond, L. (1995). Inequality and access to knowledge. In J. A. Banks & C. A. Banks (Eds.), *Handbook of research on multicultural education* (pp. 465–483). New York: Macmillan.

Darling-Hammond, L. (2004). The color line in American education: Race, resources, and student achievement. *DuBois Review, 1*(2), 213–246.

Datnow, A., & Cooper, R. (1997). Peer networks of African American students in independent schools: Affirming academic success and racial identity. *Journal of Negro Education, 66*, 56–72.

Datta, D. (1993). *Math education at its best.* Framingham, MA: The Center for Teaching/Learning of Mathematics.

Davis, S., Jenkins, G., & Hunt, R. (2002). *The pact.* New York: Penguin.

Delpit, L. D. (1988). The silenced dialogue: Power and pedagogy in educating other people's children. *Harvard Educational Review, 58*(3), 280–298

Delpit, L. D. (1995). *Other people's children.* New York: Norton.

Donelan, R., Neal, G., & Jones, D. (1994). The promise of *Brown* and the reality of academic grouping: The tracks of my tears. *Journal of Negro Education, 63*, 376–387.

Elmore, R., Peterson, P., & McCarthey, S. (1996). *Restructuring in the classroom: Teaching, learning, and school organization.* San Francisco: Jossey-Bass.

Escalante, J., & Dirmann, J. (1990). The Jaime Escalante math program. *Journal of Negro Education, 59*(3), 407–423.

Fennema, E. (1990). Teachers' attributions and beliefs about girls. *Educational Studies in Mathematics, 21*(1), 55–69.

Fennema, E., & Leder, G. (Eds.). (1990). *Mathematics and gender: Influences on teachers and students.* New York: Teachers College Press.

Ferguson, A. A. (2001). *Bad boys: Public schools in the making of Black masculinity.* Ann Arbor: University of Michigan Press.

Ferguson, R. F. (1998). Teachers' perceptions and expectations and the Black-White test score gap. In C. Jencks & M. Phillips (Eds.), *The Black–White test score gap* (pp. 273–317). Washington, DC: Brookings Institution.

Ferguson, R. F. (2002). *What doesn't meet the eye: Understanding and addressing racial disparities in high-achieving suburban schools.* Cambridge, MA: Harvard University Press.

Flores, A. (2007). Examining disparities in mathematics education: Achievement gap or opportunity gap? *The High School Journal, 91*(1), 29–42.

Flores-Gonzalez, N. (1999). Puerto Rican high achievers: An example of ethnic and academic identity compatibility. *Anthropology and Education Quarterly, 30*, 343–362.

Ford, D. Y. (1996). *Reversing underachievement among gifted Black students: Promising practices and programs.* New York: Teachers College Press.

Ford, D. Y., & Harris, J. J. III. (1999). *Multicultural gifted education*. New York: Teachers College Press.

Ford, D. Y., Harris, J. J. III, Webb, K. S., & Jones, D. L. (1994). Rejection or confirmation of racial identity: A dilemma for high-achieving Blacks? *Journal of Educational Thought, 28*, 7–33.

Fordham, S. (1988). Racelessness as a factor in Black students' school success: Pragmatic strategy or pyrrhic victory? *Harvard Educational Review, 58*, 54–85.

Fordham, S., & Ogbu, J. (1986). Black students' school success: Coping with the burden of "acting White." *The Urban Review, 18*, 176–206.

Franklin, V. P. (1990). "They rose and fell together": African American educators and community leadership, 1795–1954. *Journal of Education, 172*(3), 39–64.

Fredricks, J. A., Blumenfeld, P. C., & Paris, A. H. (2004). School engagement: Potential of the concept, state of the evidence. *Review of Educational Research, 74, 59–109*.

Fries-Britt, S. L. (*1998*). Moving beyond Black achiever isolation: Experiences of gifted Black collegians. *Journal of Higher Education, 69*(5), 556–576.

Fullilove, R. E., & Treisman, P. U. (1990). Mathematics achievement among African American undergraduates at the University of California, Berkeley: An evaluation of the mathematics workshop program. *Journal of Negro Education, 59*, 463–478.

Gándara, P. (2004). Building bridges to college. *Educational Leadership, 62*(3), 56–60.

Gándara, P., & Contreras, F. (2009). *The Latino education crisis: The consequences of failed social policies*. Cambridge, MA: Harvard University Press

Grant, P. A. (2002). Using popular films to challenge preservice teachers' beliefs about teaching in urban schools. *Urban Education, 37*(1), 77–95.

Gray, H. (1995). *Watching race: Television and the struggle for "Blackness."* Minneapolis: University of Minnesota Press.

Gilbert, S. L. (1997). The "four commonplaces of teaching": Prospective teachers' beliefs about teaching in urban schools. *The Urban Review, 29*(2), 81–96.

Givvin, K. B., Stipek, D. J., Salmon, J. M., & MacGyvers, V. L. (2001). In the eyes of the beholder: Students' and teachers' judgments of students' motivation. *Teaching and Teacher Education, 17*, 321–331.

Groulx, J. G. (2001). Changing preservice teacher perceptions of minority schools. *Urban Education, 36*(1), 60–92.

Guajardo, M. A., & Guajardo, F. J. (2004). The impact of *Brown* on the Brown of south Texas: A micropolitical perspective on the education of Mexican Americans in a south Texas community. *American Educational Research Journal, 41*, 501–526.

Gunn Morris, V., & Morris, C. L. (2000). *The price they paid: Desegregation in an African American community*. New York: Teachers College Press.

Gutiérrez, R. (1999). Advancing urban Latina/o youth in mathematics: Lessons from an effective high school mathematics department. *The Urban Review, 31*(3), 263–281.

Gutiérrez, R. (2000). Advancing African-American urban youth in mathematics: Unpacking the success of one math department. *American Journal of Education, 109*(1), 63–111.

Gutiérrez, R. (2002). Enabling the practice of mathematics teachers in context: Toward a new equity research agenda. *Mathematical Thinking and Learning, 4*(2/3), 145–187.

Gutiérrez, R. (2008). A "gap-gazing" fetish in mathematics education? Problematizing research on the achievement gap. *Journal for Research in Mathematics Education, 39*(4), 357–364.

Gutstein, E., Lipman, P., Hernandez, P., & de los Reyes, R. (1997). Culturally relevant mathematics teaching in a Mexican American context. *Journal for Research in Mathematics Education, 28,* 709–737.

Haberman, M. (1991). The pedagogy of poverty versus good teaching. *Phi Delta Kappan,* 73, 290–294.

Ham, S., & Walker, E. N. (1999). *Getting to the right algebra: The Equity 2000 Initiative in Milwaukee public schools.* New York: Manpower Demonstration Research Corporation.

Hand, V. M. (2010). The co-construction of opposition in a low-track mathematics classroom. *American Educational Research Journal, 47*(1), 97–132.

Haney, W. (1993). Minorities and testing. In L. Weis & M. Fine (Eds.), *Beyond silenced voices: Class, race, and gender in United States schools* (pp. 45–73). Albany: State University of New York Press.

Harris, D. N. (2010). How do school peers influence student educational outcomes? Theory and evidence from economics and other social sciences. *Teachers College Record, 112*(4), 1163–1197.

Hiebert, J., Carpenter, T. P., Fennema, E., Fuson, K., Wearne, D., & Murray, H. (1997). *Making sense: Teaching and learning mathematics with understanding.* Portsmouth, NH: Heinemann.

Hiebert, J., & Grouws, D. A. (2007). The effects of classroom mathematics teaching on students' learning. In F. K. Lester (Ed.), *Second handbook of research on mathematics teaching and learning: A project of the National Council of Teachers of Mathematics* (pp. 371–404). Charlotte, NC: Information Age Publishing.

Hilliard, A. (2003). No mystery: Closing the achievement gap between Africans and excellence. In T. Perry, C. Steele, & A. Hilliard (Eds.), *Young, gifted, and Black: Promoting high achievement among African-American students* (pp. 131–166). Boston: Beacon Press.

Hochschild, J. L. (1995). *Facing up to the American dream: Race, class, and the soul of the nation.* Princeton, NJ: Princeton University Press.

Horn, I. S. (2008). Turnaround students in high school mathematics: Constructing identities of competence through mathematical worlds. *Mathematical Thinking and Learning, 10,* 201–239.

Horvat, E. M., & Lewis, K. S. (2003). Reassessing the "burden of 'acting White'": The importance of peer groups in managing academic success. *Sociology of Education, 76*(4), 265–280.

Howard, T. C. (2003). "A tug of war for our minds": African American high school students' perceptions of their academic identities and college aspirations. *The High School Journal, 87,* 4–15.

Hrabowski, F. A., Maton, K. I., & Grief, G. L. (1998). *Beating the odds: Raising academically successful African American males.* New York: Oxford University Press.

Irvine, J. J. (2003). *Educating teachers for a diverse society: Seeing with the cultural eye.* New York: Teachers College Press.

Jackson, K. J. (2009). The social construction of youth and mathematics: The case of a fifth grade classroom. In D. B. Martin (Ed.), *Mathematics teaching, learning, and liberation in the lives of Black children* (pp. 175–199). New York: Routledge.

Jamar, I., & Pitts, V. R. (2005). High expectations: A "how" of achieving equitable mathematics classrooms. *The Negro Educational Review, 56*(2–3), 127–134.

Jencks, C., & Phillips, M. (Eds.). (1998). *The Black-White test score gap.* Washington, DC: Brookings Institution.

Jordan-Irvine, J. (2000). The education of children whose nightmares come both day and night. *Journal of Negro Education, 68,* 244–253.

Kea, C. D., & Bacon, E. H. (1999). Journal reflections of preservice education students on multicultural experiences. *Action in Teacher Education, 21*(2), 34–50.

Ladson-Billings, G. (1995). Toward a theory of culturally relevant pedagogy. *American Educational Research Journal, 32*(3), 465–491.

Ladson-Billings, G. (1997). It just doesn't add up: African American students' mathematics achievement. *Journal for Research in Mathematics Education, 28*(6), 697–708.

Ladson-Billings, G. (2006). From the achievement gap to the education debt: Understanding achievement in U.S. schools. *Educational Researcher, 35*(7), 3–12.

Ladson-Billings, G. (2007). Pushing past the achievement gap: An essay on the language of deficit. *The Journal of Negro Education, 76*(3), 316–323.

Lanier, H. K. (2010). What Jaime Escalante taught us that Hollywood left out: Remembering America's favorite math teacher. *Education Week, 29*(29), 32.

Lappan, G. T., & Wanko, J. J. (2003). The changing roles and priorities of the federal government in mathematics education in the United States. In G. M. A. Stanic & J. Kilpatrick (Eds.), *A history of school mathematics* (Vol. 2, pp. 897–930). Reston, VA.: National Council of Teachers of Mathematics.

Leonard, J. (2008). *Culturally specific pedagogy in the mathematics classroom.* New York: Routledge.

Lincoln, Y. S., & Guba, E. G. (1985). *Naturalistic inquiry.* Beverly Hills, CA: Sage.

Lomos, C., Hofman, R. H., & Bosker, R. J. (2011). Professional communities and student achievement—A meta-analysis. *School Effectiveness and School Improvement, 22*(2), 121–148.

Loveless, T. (1999). *The tracking and ability grouping debate.* Washington: The Thomas Fordham Foundation.

Lubienski, S. T. (1996). *Mathematics for all? Examining issues of class in mathematics teaching and learning* (Unpublished doctoral dissertation). Michigan State University, East Lansing, MI.

Lubienski, S. T. (2000). Problem solving as a means toward mathematics for all: An exploratory look through a class lens. *Journal for Research in Mathematics Education, 31*(4), 454–482.

Lubienski, S. T. (2004). Traditional or standards-based mathematics: The choices of students and parents in one district. *Journal of Curriculum and Supervision, 19*(4), 338–365.

Martin, D. B. (2000). *Mathematics success and failure among African-American youth.* Mahwah, NJ: Erlbaum.

Martin, D. B. (Ed.). (2009a). *Mathematics teaching, learning, and liberation in the lives of Black children.* New York: Routledge.

Martin, D. B. (2009b). Researching race in mathematics education. *Teachers College Record, 111*(2), 295–338.

Masingila, J. O., Muthwii, S. M., & Kimani, P. M. (2011). Understanding students' out-of-school mathematics and science practice. *International Journal of Science and Mathematics Education, 9,* 89–108.

McKinney, S. E., Chappell, S., Berry, R. Q., & Hickman, B. T. (2009). An examination of the instructional practices of mathematics teachers in urban schools. *Preventing School Failure, 53*(4), 278–284.

Meek, A. (1989). On creating *ganas*: A conversation with Jaime Escalante. *Educational Leadership, 46*(5), 46–47.

Megginson, R. E. (2003). Yueh-Gin Gung and Dr. Charles Y. Hu Award to Clarence F. Stephens for distinguished service to mathematics. *American Mathematical Monthly, 110*(3), 177–180.

Mickelson, R. A. (1990). The attitude-achievement paradox among Black adolescents. *Sociology of Education, 63*, 44–61.

Middleton, J. A., & Spanias, P. A. (1999). Motivation for achievement in mathematics: Findings, generalizations, and criticisms of the research. *Journal for Research in Mathematics Education, 30*, 65–88.

Mirel, J. (1999). *The rise and fall of an urban school system: Detroit, 1907–1981.* Ann Arbor: University of Michigan Press.

Moore, J. L III. (2006). A qualitative investigation of African American males' career trajectory in engineering: Implications for teachers, school counselors, and parents. *Teachers College Record, 108*(2), 246–266.

Moreau, M. P., Mendick, H., & Epstein, D. (2009). What do GCSE students think of mathematicians? *Mathematics in School, 38*(5), 2–4.

Moreno, J. F. (1999). *The elusive quest for equality: 150 years of Chicano/Chicana education.* Cambridge, MA: Harvard University Press.

Morrell, E. (2008). *Critical literacy and urban youth: Pedagogies of access, dissent, and liberation.* New York: Routledge.

Morris, K. A., (2006). Challenging preservice teacher expectations for their students: Response. *AMTE Connections, 15*(2), 8–9.

Moses, R. P., & Cobb, C. E. (2001). *Radical equations: Math literacy and civil rights.* Boston: Beacon Press.

Nasir, N. S., Atukpawu, G., O'Connor, K., Davis, M., Wischnia, S., & Tsang, J. (2009). Wrestling with the legacy of stereotypes: Being African American in math class. In D. B. Martin (Ed.), *Mathematics teaching, learning, and liberation in the lives of Black children* (pp. 231–248). New York: Routledge.

National Center for Education Statistics (NCES). (2007). *The condition of education 2007.* Washington, DC: Department of Education.

National Center for Education Statistics (NCES). (2009). *NAEP 2008 trends in American progress.* Washington, DC: Department of Education.

National Center for Education Statistics (NCES). (2010). *Digest of education statistics.* Washington, DC: Department of Education.

National Council of Teachers of Mathematics (NCTM). (1989). *Curriculum and evaluation standards.* Reston, VA: Author.

National Council of Teachers of Mathematics (NCTM). (2000). *Principles and standards.* Reston, VA: Author.

National Mathematics Advisory Panel. (2008). *Foundations for success: The final report of the National Mathematics Advisory Panel.* Washington, DC: U.S. Department of Education.

National Research Council (2000). *How people learn: Brain, mind, experience, and school.* Washington, DC: The National Academies Press.

Noddings, N. (1995). Teaching themes of caring. *Education Digest, 61*(3), 24–28.

Noguera, P. A. (2008). *The trouble with Black boys: . . . And other reflections on race, equity, and the future of public education.* San Francisco: Jossey-Bass.

Oakes, J. (1985). *Keeping track: How schools structure inequality.* New Haven, CT: Yale University Press.

Oakes, J. (1990) *Multiplying inequalities: The effects of race, social class, and tracking on opportunities to learn mathematics and science.* Santa Monica, CA: Rand Corporation.

Oakes, J. (1995). Two cities' tracking and within-school segregation. *Teachers College Record, 96,* 681–690.

Ogbu, J. U. (1978). *Minority education and caste: The American system in cross-cultural perspective.* New York: Academic Press

Ogbu, J. U. (1987). Variability in minority student performance: A problem in search of an explanation. *Anthropology and Education Quarterly, 18,* 312–334.

Ogbu, J. U. (1988). Class stratification, racial stratification, and schooling. In L. Weiss (Ed.), *Class, race, and gender in American education* (pp. 76–86). Albany: State University of New York Press.

Ogbu, J. U. (2003). *Black students in an affluent suburb: A study of academic disengagement.* Mahwah, NJ: Erlbaum.

Ogbu, J. U., & Davis, A. (2003). *Black American students in an affluent suburb: A study of academic disengagement.* Mahwah, NJ: Erlbaum.

Orfield, G., & Yun, J. T. (1999). *Resegregation in American schools.* Cambridge, MA: Harvard Civil Rights Project.

Osborne, J. W. (1997). Race and academic disidentification. *Journal of Educational Psychology, 89*(4), 728–735.

Perry, T., Steele, C., & Hilliard, A. G. (2003). *Young, gifted, and Black: Promoting high achievement among African American students.* Boston: Beacon Press.

Philipp, R. A. (2007). Mathematics teachers' beliefs and affect. In F. K. Lester (Ed.), *Second handbook of research on mathematics teaching and learning* (pp. 257–315). Charlotte, NC: Information Age Publishing and National Council of Teachers of Mathematics.

Picker, S. W., & Berry, J. S. (2001). Your students' images of mathematicians and mathematics. *Mathematics Teaching in the Middle School, 7,* 202–208.

Pimentel, C. (2010). Critical race talk in teacher education through movie analysis. *Multicultural Education, 17*(3), 51–56.

Polite, V. C. (1994). The method in the madness: African American males, avoidance schooling, and chaos theory. *Journal of Negro Education, 63,* 588–601.

Polite, V. C., & Davis, J. E. (Eds.). (1999). *African American males in school and society: Policy and practice for effective education.* New York: Teachers College Press.

Raymond, A. M. (1997). Inconsistency between a beginning elementary teachers' mathematics belief and practices. *Journal for Research on Mathematics Education, 28*(5), 550–576.

Reyes, L. H., & Stanic, G. M. (1988). Race, sex, socioeconomic status, and mathematics. *Journal for Research in Mathematics Education, 19,* 26–43.

Riegle-Crumb, C. (2006). The path through math: Course sequences and academic performance at the intersection of race-ethnicity and gender. *American Journal of Education, 113,* 101–122.

Robinson, D., Schofield, J. W., & Steers-Wentzell, K. L. (2005). Peer and cross-age tutoring in math: Outcomes and their design implications. *Educational Psychology Review, 17*(4), 327–362.

Rodriguez, L. F. (2008). Struggling to recognize their existence: Examining student-adult relationships in the urban high school context. *The Urban Review, 40,* 436–453.

Romo, H. D., & Falbo, T. (1996). *Latino high school graduation: Defying the odds*. Austin: University of Texas Press.

Rushton, S. P. (2004). Using narrative inquiry to understand a student-teacher's practical knowledge while teaching in an inner-city school. *The Urban Review, 36*(1), 61–79.

Schmidt, W., Houang, R., & Cogan, L. (2002). A coherent curriculum: The case of mathematics. *American Educator, 26*(2), 10–26.

Schoenfeld, A. H. (1985). *Mathematical problem solving*. Orlando, FL: Academic Press.

Schoenfeld, A. H. (1988). When good teaching leads to bad results: The disasters of "well-taught" mathematics courses. *Educational Psychologist, 23*(2), 145–166.

Schoenfeld, A. H. (2002). Making mathematics work for all children: Issues of standards, testing, and equity. *Educational Researcher, 31*(1), 13–25.

Secada, W. G. (1992). Race, ethnicity, social class, language, and achievement in mathematics. In D. A. Grouws (Ed.), *Handbook of research on mathematics teaching and learning* (pp. 623–660). New York: Macmillan.

Siddle-Walker, V. (1996). *Their highest potential: An American school community in the segregated South*. Chapel Hill: University of North Carolina Press.

Solorzano, D. (1992). An exploratory analysis of the effects of race, class, and gender on student and parent mobility aspirations. *Journal of Negro Education, 61*, 30–44.

Spencer, J. (2009). Identity at the crossroads: Understanding the practices and forces that shape African American success and struggle in mathematics. In D. B. Martin (Ed.), *Mathematics teaching, learning, and liberation in the lives of black children* (pp. 200–230). New York: Routledge.

Stanton-Salazar, R. D. (2001). *Manufacturing hope and despair: The school and kin support networks of U.S.-Mexican youth*. New York: Teachers College Press.

Steele, C. M. (1997). A threat in the air: How stereotypes shape intellectual identity and performance. *American Psychologist, 52*, 613–629.

Steinberg, L., Brown, B. B., & Dornbusch, S. M. (1996). *Beyond the classroom: Why school reform has failed and what parents need to do*. New York: Simon & Schuster.

Steinberg, L., Dornbusch, S. M., & Brown, B. B. (1992). Ethnic differences in adolescent achievement: An ecological perspective. *American Psychologist, 47*, 723–729.

Stern, A. L., & McCrocklin, E. (2006). *What works best in science & mathematics education reform*. Washington, DC: Potomac Communications Group.

Stiff, L. V., & Harvey, W. B. (1988). On the education of black children in mathematics. *Journal of Black Studies, 19*, 190–203.

Stigler, J. W., & Hiebert, J. (1999). *The Teaching Gap*. New York: Free Press.

Stinson, D. W. (2006). African-American male adolescents, schooling (and mathematics): Deficiency, rejection, and achievement. *Review of Educaitonal Research, 76*(4), 477–506.

Stipek, D. J., Givvin, K. B., Salmon, J. M., & MacGyvers, V. L. (2001). Teachers' beliefs and practices related to mathematics instruction. *Teaching and Teacher Education, 17*(4), 213–226.

Strutchens, M., Lubienski, S. T., McGraw, R., & Westbrook, S. K. (2004). NAEP findings regarding race and ethnicity: Students' performance, school experiences, attitudes and beliefs, and family influences. In P. Kloosterman & F. Lester (Eds.), *Results and interpretations of the 1990-2000 mathematics assessments of the National Assessment of Educational Progress* (pp. 269–305). Reston, VA: National Council of Teachers of Mathematics.

Students from Bronx Leadership Academy 2, O'Grady, S., Ferrales, K., & Cushman, K. (2008). *SAT Bronx: Do you know what Bronx kids know?* Providence, RI: Next Generation Press.

Suskind, R. (1998). *A hope in the unseen: An American odyssey from the inner city to the Ivy League.* New York: Broadway Books.

Tanton, J. (2006). Math circles and olympiads. MSRI asks: Is the US Coming of Age? *Notices of the American Mathematical Society, 53*(2), 200–205.

Tate, W. F. (1994). From inner city to ivory tower: Does my voice matter in the academy? *Urban Education, 29*, 245–269.

Tate, W. F. (1995). Returning to the root: A culturally relevant approach to mathematics pedagogy. *Theory into Practice, 34*(3), 166–173.

Tate, W. F. (1997). Race-ethnicity, SES, gender and language proficiency trends in mathematics achievement. *Journal for Research in Mathematics Education, 28*(6), 652–679.

Tatum, B. D. (1997). *"Why are all the Black kids sitting together in the cafeteria?" and other conversations about race.* New York: Basic Books.

Thirunarayanan, M. O. (2004). The "significantly worse" phenomenon: A study of student achievement in different content areas by school location. *Education and Urban Society, 36*(4), 467–481.

Treisman, U. (1992). Studying students studying calculus: A look at the lives of minority mathematics students in college. *College Mathematics Journal, 23*, 362–372.

Uekawa, K., Borman, K., & Lee, R. (2007). Student engagement in U.S. urban high school mathematics and science classrooms: Findings on social organization, race, and ethnicity. *The Urban Review, 39*(1), 1–43.

U.S. Department of Education. (1983). *A nation at risk.* Washington, DC: U.S. Government Printing Office.

Useem, E. L. (1990). You're good, but not good enough: Tracking students out of advanced mathematics. *American Educator, 14*, 24–46.

Useem, E. L. (1992). Getting on the fast track in mathematics: School organizational influences on math track assignment. *American Journal of Education, 100*(3), 325–353.

Valenzuela, A. (1999). *Subtractive schooling: U.S.-Mexican youth and the politics of caring.* Albany: State University of New York Press.

Valli, L. (1995). The dilemma of race: Learning to be color blind and color conscious. *Journal of Teacher Education, 46*(2), 120–129.

Vetter, B. M. (1994). The next generation of science and engineers: Who's in the pipeline? In W. Pearson & A. Fetcher (Eds.), *Who will do science? Educating the next generation.* Baltimore: Johns Hopkins University Press.

Vogler, K. E., & Burton, M. (2010). Mathematics teachers' instructional practices in an era of high-stakes testing. *School Science and Mathematics, 110*(5), 247–261.

Walker, E. N. (2001). *On time and off track? Advanced mathematics course-taking among high school students.* Unpublished doctoral dissertation, Harvard University, Cambridge, MA.

Walker, E. N. (2003). Who can do mathematics? In B. Vogeli & A. Karp (Eds.), *Activating mathematical talent* (pp. 15–27). Boston: Houghton Mifflin and National Council of Supervisors of Mathematics.

Walker, E. N. (2006). Urban high school students' academic communities and their effects on mathematics success. *American Educational Research Journal, 43*(1), 43–73.

Walker, E. N. (2009a). More than test scores: How teachers' classroom practice contributes to and what student work reveals about Black students' mathematics performance and understanding. In D. B. Martin (Ed.), *Mathematics teaching, learning, and liberation in the lives of Black children* (pp. 145–171). New York: Routledge.

Walker, E. N. (2009b). (Un)limited opportunity: The experiences of mathematically talented Black American students. In M. Tzekaki, M. Kaldrimidou, & H. Sakonidis (Eds.), *Proceedings of the 33rd Conference of the International Group for the Psychology of Mathematics Education: In Search for Theories in Mathematics Education* (Vol. 1, pp. 1-199–1-203). Thessaloniki, Greece: PME.

Walker, E. N. (2010). *Reimagining mathematical spaces: Cultivating mathematics identities in and out of school and in between.* Working paper, Teachers College, Columbia University, New York, NY.

Walker, E. N. (2011). Supporting giftedness: Historical and contemporary contexts for mentoring within Black mathematicians' academic communities. *Canadian Journal for Science, Mathematics, and Technology Education, 11*(1), 19–28.

Walker, E. N., & McCoy, L. P. (1997). Student voices: African Americans and mathematics. In J. Trentacosta & M. J. Kenney (Eds.), *Multicultural and gender equity in the mathematics classroom: The gift of diversity* (pp. 34–45). Reston, VA: National Council of Teachers of Mathematics.

Webb, N. M., & Mastergeorge, A. M. (2003). The development of students' helping behavior and learning in peer-directed small groups. *Cognition and Instruction, 21*(4), 361–428.

Weis, L., & Fine, M. (2000). *Construction sites: Excavating race, class, and gender among urban youth.* New York: Teachers College Press.

Werkema, R. D., & Case, R. (2005). Calculus as a catalyst: The transformation of an inner-city high school in Boston. *Urban Education, 40,* 497–520.

Wiggan, G. (2007). Race, school achievement, and educational inequality: Toward a student-based inquiry perspective. *Review of Educational Research, 77*(3), 310–333.

Woolley, M. E., Strutchens, M. E., Gilbert, M. C., & Martin, W. G. (2010). Mathematics success of Black middle school students: Direct and indirect effects of teacher expectations and reform practices. *Negro Educational Review, 61*(1–4), 41–59.

Yan, W. (1999). Successful African American students: The role of parental involvement. *Journal of Negro Education, 68,* 5–22.

Yonezawa, S., Wells, A. S., & Serna, I. (2002). Choosing tracks: "Freedom of choice" in detracking schools. *American Educational Research Journal, 39,* 37–67.

Zollman, A., Smith, C. M., & Reisdorf, P. (2011). Identity development: Critical components for self-regulated learning in mathematics. In D. Brahier (Ed.), *Motivation and disposition: Pathways to learning mathematics—Seventy-third yearbook* (43–53). Reston, VA: National Council of Teachers of Mathematics.

Index

A Different World (TV show)
  cultural impact of, 11
  mathematics portrayal in, 10
A Nation at Risk, 53, 54, 56
Academic achievement. *See also* Student
  achievement, in mathematics
  parental support for, 38
  peer influences on, 33–39
  peer responses to, 46–47, 46t–47t
  peer support for, 36
  self-esteem and, 37
  social sanctions for, 36–37
Academic behaviors, modeling of, 38, 48
Access, to mathematics opportunities
  for Black students, 16, 64, 112
  college-preparatory courses, 14–15, 64
  for Latino/a students, 16, 64, 112
  STEM programs, 60
Accountability, standardized assessments
  and, 76
Achievement. *See* Academic
  achievement; Student achievement,
  in mathematics
Achievement gap, 12–17, 51. *See also*
  Performance gaps
"Acting White" hypothesis, 35
  gender differences and, 37
Administrators, implications of
  reconceptualizing mathematics
  teaching and learning for, 114–115
Adolescence, peer *vs.* parental influence
  during, 34–35
Advanced courses, 15
  avoiding, reasons for, 35–36
  in calculus, 61
  participation in, 26

Advisors, for peer tutors
  anecdotal evidence from, 95
  experiences of, 94–95
  recruitment and training of, 93
  and tutor–advisor collaboration, 99–106
African American students. *See* Black
  students
Ainsworth-Darnell, J. W., 35
"Algebra for All" initiatives, 54
Algebra Project, 82–84, 87
Allexsaht-Snider, M., 64
American Association of University
  Women (AAUW), 25
Anderson, J., 29, 30
Anxiety, mathematics, 73–74
Anyon, J., 13, 58
AP (advanced placement) courses. *See*
  Advanced courses
Appelbaum, P. M., 10, 11, 16
Asian students
  in college preparatory courses, 14–15
  drop-out rates, 129n3
  performance in mathematics, 13–14
  success in mathematics, reasons for, 49
Atukpawu, G., 12, 15–16
Auerbach, S., 30, 33, 35, 59
Azmitia, M., 30, 35, 38

Bacon, E. H., 19
Bakari, R., 19, 20
Bard College, Summer Program in
  Mathematical Problem Solving,
  118–119
Beliefs. *See* Teachers' beliefs
Bell, L. A., 19, 21, 58
Bell, T., 53

Bempechat, J., 36, 45
Berry, J. S., 9
Berry, R. Q., 13, 19, 37, 74
Black community
    parents in, 30
    support for education and mathematics
        learning, 29–30
Black students, 1
    access to mathematics opportunities,
        16, 64
    in college preparatory courses, 14–15
    drop-out rates, 15–16
    in lower-level courses,
        overrepresentation of, 57
    and mathematics, shattering stereotypes
        of, 10–11
    media portrayal of, 7–12
    peer support for, 34
    performance in mathematics, 13–14
    racelessness stance adopted by, 36
    teacher encouragement and, 66
Blumenfeld, P. C., 73
Boaler, J., 8, 61, 62, 72, 101
Bol, L., 19
Books, mathematics portrayal in, 9
Borman, K., 75
Bosker, R. J., 61
Breitborde, M., 19, 58
Brown, B. B., 30, 34, 35, 36
Brown v. Board of Education, 30, 56
Burris, C. C., 58, 115
Burton, L., 10
Burton, M., 76

Cahnmann, M. S., 63, 75
Calculus
    college-level courses in, 61
    course offerings, examples of, 59–60
    success/failure rates in, 49
Cammarota, J., 35, 38
Camouflaging, 38
Campbell, P., 26
Caring, ethic of, 67
Carpenter, T. P., 90
Case, R., 15, 60, 61, 115
Castro, A. J., 19
Catsambis, S., 25, 26

Chappell, S., 13, 74
Chavira, G., 30, 35, 38
Checkley, K., 83
Cirillo, M., 10, 72
Civil Rights Project, 56
Cobb, C. E., 8, 58, 83, 90, 115
Cobb, P., 18, 101
Code-switching, 38
Cogan, L., 78
Collaboration, on mathematics work,
    43–44
    peer tutoring program and. See Peer
        Tutoring Collaborative
Collaborative space
    creating, for mathematics, 106–107
    development of, 118, 119
    logistical and organizational concerns
        about, 111
College Board, Equity 2000 program of, 54
College-level courses
    STEM programs, key concepts of, 118
    summer opportunities for, 61
College-preparatory courses, in
    mathematics
    access to, 14–15, 64
    student body characteristics and, 64
    university initiatives, 49
Common Core State Standards
    adoption of, mathematics performance
        and, 116
    origins of, 56
Community. See Educational communities;
    Mathematics communities
Competition, between students, 45–46
Conceptual understanding, practices
    promoting, 65
Conchas, G. Q., 30, 35, 36
Content, mathematics engagement and,
    74, 87
Contreras, F., 18
Cook, P. J., 35
Cooper, C. R., 30, 35, 38
Cooper, C. W., 37, 59
Cooper, R., 30, 32, 36, 38
Cooper, R. G., 30, 35, 38
Cooperative learning groups, 75
Corbett, D., 66, 67, 77

Corbin, K., 37
Cosby, Bill, 11
The Cosby Show (TV show), cultural
    impact of, 11
Counselors, importance of, 60
Cromidas, R., 118
Crosnoe, R., 52
Cultural/linguistic background, of
    students, impact on mathematics
    teaching and learning, 18, 75, 86
Curriculum and Evaluation Standards
    (NCTM), 54, 55, 56
Cushman, K., 71

Dangerous Minds (movie), 65
Darling-Hammond, L., 14, 57, 64
Datnow, A., 30, 32, 36, 38
Datta, D., 49
Davis, A., 36, 57
Davis, J. E., 37
Davis, M., 12, 15–16
Davis, S., 34, 44
de los Reyes, R., 66
Decision-making, in schools, 115–116
Delpit, L. D., 59, 66, 67, 75
Dirmann, J., 10, 11, 60, 115
Donelan, R., 57, 58
Dornbusch, S. M., 30, 34, 35, 36
Downey, D. B., 35
Drop-out rates, 15

Education
    community support for, 29–30
    systemic inequities in, 57–58
Educational communities
    Black and Latino/a, 29–30
    mathematics engagement and, 86
    peer tutoring programs and, 113–114.
        See also Peer Tutoring Collaborative
Educational Testing Service (ETS), 10, 11
Effective teachers. *See* Good teachers
Elementary and Secondary Education Act
    of 1965, 53
Elmore, R., 56
Engagement in mathematics
    classroom measurements and
        experiences, 75, 113

components of, 73
    enrichment programs and, 82–86
    framework for, 86–87
    mathematicians' experiences described,
        77–78
    and nonengagement experiences
        described, 77–78
    parental support for, 59
    reform mathematics and, 76
    student experiences described, 77–78,
        113
English proficiency, word problems and, 72
Enrichment programs
    Algebra Project, 82–84
    funding for, 87
    math circles, 85–86, 87
    during summer recess. *See* Summer
        enrichment programs
    sustainability of, 111
Epstein, D., 9, 10
Equity 2000 program, 54
Escalante, J., 10, 11, 16, 60, 115
Ethic of caring, 67
Everybody Counts, 54
Everyday activities, of students
    achievement related to, 72–73
    validation of, 86
Expectations
    of parents of Lowell High School
        students, 31–32, 45
    of peers, 39–40
    by teacher for students of color, 17–18,
        57, 63
Experience Sampling Method (ESM), 75

Falbo, T., 33
Family Matters (TV show), 12
Family support, for mathematics learning,
    31–33
Fennema, E., 25, 26, 74, 90
Ferguson, A. A., 7, 37, 38
Ferguson, R. F., 4, 18, 66
Ferrales, K., 71
Film, mathematics portrayal in, 9, 65–66
Fine, M., 18
Fliers, for student recruitment, 92, 125
Flores, A., 16, 17

Flores-Gonzalez, N., 34, 36, 38, 39
Ford, D. Y., 14, 35, 38
Fordham, S., 35, 36, 38
Franklin, V. P., 30
Fredricks, J. A., 73
Fries-Britt, S. L., 50
Fullilove, R. E., 49, 75, 90
Funding
    for enrichment programs, 87
    by National Science Foundation, 53, 54
Fuson, K., 90

Gándara, P., 18, 64
Garfield High School (East Los Angeles)
    Escalante myth and, 16
    student achievement at, 10–11, 16
Gender differences
    in academic achievement, 37
    in attitudes toward mathematics, 25–26,
        28, 117
    in course-taking behavior, 26–27
    in perceptions of peer group support,
        47–48
    in sciences, 9
Gilbert, M. C., 75
Gilbert, S. L., 19, 58
Givvin, K. B., 74, 76
Good teachers
    characteristics of, 66
    student descriptions of, 66, 67–70
    studies on, 65–66
Grant, P. A., 7, 10, 17, 65, 77
Gray, H., 11
Greeno, J., 8
Grief, G. L., 49, 60, 90
Groulx, J. G., 19, 90, 108
Groups/Grouping
    for cooperative learning, 75
    school policies on, examples of, 59–60
Grouws, D. A., 65
Guajardo, F. J., 30
Guajardo, M. A., 30
Guba, E. G., 122
Gullat, Y., 30, 35, 38
Gunn Morris, V., 29–30
Gutiérrez, R., 15, 17, 18, 61, 64, 76, 115
Gutstein, E., 66

Haberman, M., 58, 115
Ham, S., 54
Hand, V, M., 18
Haney, W., 57, 58
Harris, D. N., 36, 57
Harris, J. J. III, 14, 35, 38
Hart, L. E., 64
Harvey, W. B., 15
Head Start program, 53
Herbel-Eisenmann, B., 10, 72
Hernandez, P., 66
Heubert, J., 58, 115
Hickman, B. T., 13, 74
Hiebert, J., 24, 65, 90
High achievers
    ambivalence toward, 41–43
    help and encouragement provided by,
        43–45
    influences on, 48–49
    peer responses to, 40–46
    as peer tutors, 90. See also Peer tutors
High school. See Urban schools
High-track mathematics courses, 58
Hilliard, A. III, 17, 30, 33, 38, 90
Hochschild, J. L., 20, 38
Hodge, L. L., 18, 101
Hofman, R. H., 61
Honor students, 15
Horn, I. S., 18
Horvat, E. M., 35, 36, 37, 38, 39
Houang, R., 78
Howard, T. C., 30
Hrabowski, F. A., 49, 60, 90
Hunt, R., 34, 44

Irvine, J.J., 90

Jackson, K. J., 18
Jamar, I., 58, 59
Jencks, C., 12, 13, 129n1
Jenkins, G., 34, 44
Jones, D., 57, 58
Jones, D. L., 35, 38
Jordan-Irvine, J., 30

Kea, C. D., 19
Kimani, P. M., 72

Ladson-Billings, G., 13, 16, 17, 51, 58, 59, 66, 67, 68
Lanier, H. K., 10, 11
Lappan, G.T., 53
Latino/a community
  parents in, 30
  support for education and mathematics learning, 29–30
Latino/a students, 1
  access to mathematics opportunities, 16, 64
  in college preparatory courses, 14–15
  drop-out rates, 15–16, 129n3
  in lower-level courses, overrepresentation of, 57
  and mathematics, shattering stereotypes of, 10–11
  media portrayal of, 7–12
  performance in mathematics, 13–14
Lectures, 75
Leder, G., 25, 26
Lee, R., 75
Leonard, J., 66
Levin, H., 58, 115
Lewis, K. S., 35, 36, 37, 38, 39
Lincoln, Y. S., 122
Lipman, P., 66
Literacy level, word problems and, 72
Lomos, C., 61
Loveless, T., 57
Low achievers, standardized assessments and, 117
Lowell High School, New York City, 1
  engagement experiences at, 78–82
  family support at, 31–33
  nonengagement experiences at, 77–78
  parental expectations and, 31–32, 45
  peer tutoring model at. *See* Peer Tutoring Collaborative
  research project methodology, 121–123
  student attitudes and behaviors toward mathematics, 27–28
  student characteristics and performance, 21–24, 27, 121
  student peer groups at, 39–48, 64
  teacher evaluations by students, 67–70
  teacher qualifications at, 121

Lower-track mathematics courses
  students of color overrepresented in, 57
  teacher assignment to, 58
Lubienski, S. T., 13, 18, 25, 34, 73, 75
Ludwig, J., 35

MacGyvers, V. L., 74, 76
Map, student, peer tutoring and, 124
Martin, D. B., 16, 17, 18, 35, 76, 118
Martin, W. G., 75
Masingila, J. O., 72
Mastergeorge, A. M., 90
Math circles, 85–86, 87
Mathematical Problem Solving (Summer Program at Bard College), 118–119
Mathematicians
  on early out-of-school mathematics experience, 83–85, 116
  on engagement experiences, 80–82
  stereotypes of, 9
  teacher characterizations of, 10
  TV show characters as, 11–12
Mathematics
  college preparatory courses, access to, 14–15, 64
  formal, problem-solving *vs.*, 72
  high achievement in, peer responses to, 46–47, 46t–47t. *See also* High achievers
  meaningful, 72, 119
  media portrayal of, 8–12
  and peer tutoring outcomes, 94–95
  popular perception of, 8–9
  racial-ethnic participation/performance in, 13–17, 26–27, 129n4
  reform-based vs.traditional approach, choosing between, 74
  socialization issues and. *See* Mathematics socialization
  status in high school, 1–2, 72
  student attitudes and behaviors toward, 27–28
  teachers' beliefs about (Lowell High School findings), 23
Mathematics anxiety, 73–74
Mathematics classrooms
  student experiences in, 75, 113

Mathematics classrooms *(continued)*
  student perceptions of, 75
  teacher educators and, 114–115
  teaching approaches in, 74–77, 113
Mathematics communities. *See also* Peer
    Tutoring Collaborative
  benefits of, 2
  principles for building, 2–6
Mathematics departments, effective
    structure and practice characteristics,
    61–62
Mathematics reform
  departmental policies and, 61–62
  impact of, 56
  meaningful, problems implementing, 72
  student engagement and, 76
  teacher-student interaction and, 75–76
  teachers' beliefs about (Lowell High
    School findings), 21–24
Mathematics socialization
  engagement and, 74, 87
  spaces for, 119
Mathematics teaching and learning
  classroom activities, ethnic/racial
    preferences for, 75
  community support for, 29–30
  constructive/deconstructive classroom
    practices, 18–19
  creating collaborative space for, 106–107
  effective, components of, 65–67
  family support for, 31–33
  film and television depiction of, 65
  impact of cultural and linguistic
    backgrounds on, 18, 75
  inequalities in, 13–14, 129n2
  linking with out-of-school mathematics,
    82–86
  peer group support for, 32–33
  reconceptualizing. *See* Reconceptualizing
    mathematics teaching and learning,
    implications of
  reform perspectives, 2, 22
  teachers' beliefs about, 17–24, 62–63
Mathematics Workshop Program
    (University of California at Berkeley),
    49
Maton, K. I., 49, 60, 90

McCarthey, S., 56
McCoy, L. P., 35
McCrocklin, E., 54
McGraw, R., 13, 25, 34
McKinney, S. E., 13, 74
Media
  Black and Latino/a student portrayal
    in, 7–8
  Black male academic development and,
    38
  mathematics portrayal in, 8–12
Meek, A., 10
Megginson, R. E., 49
Mendick, H., 9, 10
Mentors. *See* Advisors, for peer tutors
Meyerhoff Scholars Program, 49
Mickelson, R. A., 25
Middleton, J. A., 45
Milner, H. R., 64
Mirel, J., 30
Moore, J. L. III, 37, 60, 64
Moreau, M. P., 9, 10
Moreno, J. F., 30
Morgan State University, mathematics
    initiative at, 49–50
Morrell, E., 7
Morris, C. L., 29–30
Morris, K. A., 64, 83
Moses, R. P., 8, 58, 83, 90, 115
Motivation of students
  research studies on, 73–74
  teachers' beliefs about, 20–21
Murray, H., 90
Muthwii, S. M., 72

Nasir, N. S., 12, 15–16
National Assessment of Education Progress
    (NAEP), 13, 15, 24, 25
National Center for Education Statistics
    (NCES), 1, 13, 15, 24, 26
National Council of Teachers of
    Mathematics (NCTM), 19, 54, 55, 74,
    90, 110
National Education Longitudinal Study of
    1988 (NELS 88), 35
National Mathematics Advisory Panel, 74,
    76

National Research Council (NRC), 54, 74
National Science Foundation (NSF)
    funding by, 53, 54
    mathematics reform and, 74
Native American students
    in college preparatory courses, 14–15
    drop-out rates, 15–16, 129n3
    performance in mathematics, 13–14
NCLB. *See* No Child Left Behind Act of
    2001
Neal, G., 57, 58
"Near peers," 48
    as role models, 32
    support provided by, 34
No Child Left Behind Act of 2001, 53–54,
    55, 116
Noddings, N., 67
Noguera, P. A., 37, 38

Oakes, J., 14, 15, 57, 58, 64, 129n2
O'Connor, K., 12, 15–16
Ogbu, J. U., 35, 36, 37, 38, 57
O'Grady, S., 71
187 (movie), 65
Orfield, G., 30
Osborne, J. W., 34, 37
Out-of-school mathematics
    linking with in-school mathematics
        learning, 82–86, 87, 116
    mathematics engagement and, 86
    student practices and behaviors, 72–73
Outcomes, of peer tutoring, 94–95
Overachievers, depiction of, 9

Parents
    academic achievement encouraged by, 38
    counterexamples provided by, 33
    educational values of, 19–20
    involvement in school decisions, 59
    of Lowell High School students,
        expectations of, 31–32, 45
    popular perceptions of, 30
Paris, A. H., 73
"Pedagogy of poverty," 58
Peer groups. *See also* "Near peers"
    academic achievement influenced by,
        33–39

gender differences in perceptions of, 47–48
    at Lowell High School, 39–48, 64
    support for mathematics learning,
        32–33, 36, 39–48, 64
Peer pressure, academic achievement and,
    36–37
Peer support, reciprocal nature of, 44
Peer tutoring, 49–50
    as community support, 113–114
    distractions during, 106
    recruitment materials, 125
    sample materials, 126–127
    session organization in, 93–94
    tips for, 128
Peer Tutoring Collaborative
    approaches in, 88–89
    components of, 91–94
    development of, 6, 88
    methodological notes, 121–123
    origins of, 89
    outcomes of, 94–95, 129n2
    rationale for, 89–90
    recruitment of student participants,
        92–93
    reinstatement of, 110–111
    student map exercise, 124
Peer tutors
    advisors for, recruitment and training
        of, 93
    anecdotal evidence from, 95
    background information of, 91–92
    confidence levels among, 107
    high-achieving students as, 90, 122–123
    recruitment of, 91, 124
    training materials for, 91, 126–128
    and tutor-advisor collaboration, 99–106
    and tutor-tutee interactions, 96–99, 106
Performance gaps
    closing, investment in, 115
    interpretive positions on, 17
    peer group influences and, 34–35
    racial-ethnic participation in
        mathematics and, 14–17
    reasons for, 13–14
    teachers' beliefs about, 19–20
    widening of, 13, 129n1
Perry, T., 30, 33, 38

Peterson, P., 56
Philipp, R. A., 17
Phillips, M., 12, 13, 129n1
Picker, S. W., 9
Pimentel, C., 7, 10–11
Pitts, V. R., 58, 59
Policies and practices, impact on student
    achievement
    Black male academic development and, 38
    at department level, 61–62
    examples of, 51–53
    helpful, at Lowell High School, 70
    at national level, 53–56
    reconceptualizing mathematics teaching
        and learning and, 114–115
    at school level, 56–60, 62–64
Policy makers, implications of
    reconceptualizing mathematics
    teaching and learning for, 114–115
Polite, V. C., 35, 37, 59, 66
Poverty, pedagogy of, 58
Preservice teachers
    as mentors/advisors. *See* Advisors, for
        peer tutors
    student stereotyping by, 114–115
    urban school perceptions of, 19–20,
        107–110
Principles and Standards, 54, 55, 56
Problem-solving, formal mathematics *vs.*, 72
Procedural fluency, development of, 65
Program for International Student
    Assessment (PISA), 24
Project-based activities, English proficiency
    and, 72
Pruitt, R., 37

Racelessness, Black students and, 36
Raymond, A. M., 23
Recognition, importance of, 67
Reconceptualizing mathematics teaching
    and learning, implications of
    for access opportunities, 112
    for administrators and policy makers,
        115–116
    for student achievement, 117–119
    for teacher educators, 114–115
    for teachers, 112–114

Recruitment materials, for peer tutoring,
    92, 125
Reisdorf, P., 74
Remillard, J. T., 63, 75
Respect, importance of, 68
Reyes, L. H., 63
Riegle-Crumb, C., 14, 26, 27, 37
Rigor, student validation and, 86
Robinson, D., 88
Rodriguez, L. F., 67
Role models, family members as, 32–33, 48
Romo, H. D., 33
Rushton, S. P., 19, 90

Salmon, J. M., 74, 76
Salt (movie), 9
SAT Bronx project, 71–72
Schmidt, W., 78
Schoenfeld, A. H., 53, 54, 56, 72
Schofield, J. W., 88
School counselors, importance of, 60
Schools. *See* Urban schools
Science, technology, engineering, and
    mathematics (STEM) programs
    access to, 60
    key concepts of, 118
Seatwork, 75
Secada, W. G., 14, 57, 58, 76
Self-esteem, academic achievement and, 37
Serna, I., 30, 36, 38, 64
Siddle-Walker, V., 30
Skills efficiency (procedural fluency),
    development of, 65
Smith, C. M., 74
Social sanctions, for academic
    achievement, 36–37
Socialization in mathematics
    engagement and, 74, 87
    spaces for, 119
Socioeconomic status (SES)
    detracking policies and, 58
    parental involvement in school decisions
        and, 59
    student attitudes toward mathematics
        and, 25
Solorzano, D., 38
Space. *See* Collaborative space

Spanias, P. A., 45
Spencer, J., 18
Stand and Deliver (movie), 10–11, 17, 65
  Escalente myth and, 16
Standardized assessments
  default measures in, 76
  low achievers and, 117
  test scores influencing class placement, 57
Standards, in mathematics
  adoption of, 116
  common. *See* Common Core State
    Standards
  development of, 54
Stanic, G. M., 63
Stanton-Salazar, R. D., 30, 35, 38, 44
Staples, M., 61
State University of New York (SUNY),
    mathematics initiatives at, 49
Steele, C., 30, 33, 38, 118
Steers-Wentzell, K. L., 88
Steinberg, L., 30, 34, 35, 36
STEM (science, technology, engineering,
    and mathematics) programs
  access to, 60
  key concepts of, 118
Stephens, C. F., 49–50
Stereotyping
  of mathematicians, 9–11
  of students of color, 10–11, 114–115
  of urban schools, 19–20
Stern, A. L., 54
Stiff, L. V., 15
Stigler, J. W., 24
Stinson, D. W., 37, 118
Stipek, D. J., 74, 76
Strutchens, M. E., 13, 25, 34, 75
Student achievement, in mathematics
  attitudes and, 24–26
  course-taking behavior and, 26–27
  at different school levels, 14–15
  everyday activities and, 72–73
  gaps in. *See* Performance gaps
  at Garfield High School (East Los
    Angeles), 10–11
  gender gap in, 25–26
  high. *See* High achievers
  improvement in, investment in, 115

at Lowell High School. *See* Lowell High
    School, New York City
  media portrayal of, 7–12
  networks and communities supporting.
    See Mathematics communities
  peer influences on, 33–39
  peer responses to, 46–47, 46t–47t
  peer tutoring model and. *See* Peer
    Tutoring Collaborative
  policies and practices impacting,
    examples of, 51–53
  racial-ethnic participation/performance
    in, 12–17
  reconceptualizing, implications of,
    117–119
  sample student map, 124
  successful, strategies fostering, 61–62
  support for. *See* Mathematics
    communities
  and underachievement, theory of, 35
  validation and, 86
Student map, peer tutoring and, 124
Student participants, in Peer Tutoring
    Collaborative
  anecdotal evidence from, 95
  recruitment of, sample materials, 92, 125
  and tutor-tutee interactions,
    characteristics of, 96–99, 106
Students of color. *See also* Black students;
    Latino/a students
  confidence levels of, 25
  course-taking behavior of, 26–27
  peer influence on, 34–35
  stereotyping of, 10–11, 114–115
  teacher expectations of, 17–18
Summer enrichment programs
  college-level courses, 61
  Mathematical Problem Solving (Bard
    College), 118–119
  mathematicians' experiences of, 84–85
Summers, L., 9
Supervisors, in Peer Tutoring Collaborative
  anecdotal evidence from, 95
  logistical and organizational concerns
    about, 111
  teachers as, 93
Suskind, R., 36

Tanton, J., 85, 87
Tate, W. F., 34, 37, 54–55, 56, 115, 129n2
Tatum, B. D., 36
Teacher educators, implications of
reconceptualizing mathematics
teaching and learning for, 114–115
Teacher preparation programs
implications of reconceptualizing
mathematics teaching and learning
for, 114–115
negative attitudes emulated in, 19–20
preservice perceptions and, 19
student stereotyping and, 114–115
Teacher-student interaction, 62
dynamics of, 63–64
reform-initiated practices and, 75–76
Teachers
beliefs of. *See* Teachers' beliefs
effective. *See* Good teachers
encouragement from, Black students
and, 66
identification of talented mathematics
students by, 112–114
implications of reconceptualizing
mathematics teaching and learning
for, 114–115
importance of, 60
ineffective, studies on and student
descriptions of, 65–66
instructional behaviors of, 113
preservice. *See* Preservice teachers
student evaluation of, 66
student stereotyping and, 114–115
as supervisors in Peer Tutoring
Collaborative, 93, 95
Teachers' beliefs
about how students spend their time,
47, 48t
about mathematics teaching in urban
schools, 17–21, 62–63, 64
about peer responses to academic/
mathematics success, 46–47,
46t–47t
at Lowell High School, research findings,
21–24
preservice perceptions and, 19–20,
107–110

student stereotyping and, 114–115
traditional practices *vs.* inquiry-oriented
approaches and, 76–77
Television, mathematics portrayal on, 10,
65–66
cultural impact of, 11
Thirunarayanan, M. O., 51
Time, how students spend with friends,
47, 48t
Tracking, 57–58
Training, of peer tutors, 91
sample materials, 126–128
Treisman, P. U., 49, 75, 90
Treisman, U., 49, 118
Trends in Mathematics and Science Study
(TIMSS), 74–75
Tsang, J., 12, 15–16
Tutoring. *See* Peer tutoring
Tutors. *See* Peer tutors

Uekawa, K., 75
Underachievement, in mathematics. *See*
Student achievement, in mathematics
University of California (Berkeley),
Mathematics Workshop Program at, 49
Urban schools. *See also individually named*
high schools
administrative decision-making in,
115–116
building mathematics communities in.
See Mathematics communities
counselors in, importance of, 60
implications of reconceptualizing
mathematics teaching and learning
for, 115–116
mathematics teaching in, teachers'
beliefs about, 17–21, 62–63, 64
measuring student intelligence in, 71–72
parental involvement discouraged by, 59
preservice perceptions about, 19–20,
107–110
resources and organization practices of,
56–60
status of mathematics in, 1–2, 72
student underperformance in, 51
teaching-learning paradigm in, firm and
television depiction of, 65–66

Urban Systemic Initiatives/Program, 54
Urkel, Steve (TV character), 11–12
U.S. Department of Education, 53
Useem, E. L., 14, 57, 59, 60, 115

Valenzuela, A., 30, 38, 59, 67, 68
Validation, of students, 86
Valli, L., 19, 108
Vetter, B. M., 8
Vogler, K. E., 76

Walker, E. N., 1, 14, 26, 27, 34, 35, 38, 49,
    54, 58, 62, 80, 118, 129n4
Wanko, J. J., 53
Wayne, Dwayne (TV character), 11–12
Wearne, D., 90
Webb, K. S., 35, 38
Webb, N. M., 90
Weis, L., 18
Wells, A. S., 30, 36, 38, 64

Werkema, R. D., 15, 60, 61, 115
Westbrook, S. K., 13, 25, 34
White students
    in college preparatory courses, 14–15
    drop-out rates, 129n3
    performance in mathematics, 13–14
Wiggan, G., 16
Wilson, B., 66, 67, 77
Wischnia, S., 12, 15–16
Women, in sciences, 9
Woolley, M. E., 75
Word problems, English proficiency and, 72

Yan, W., 59
Yonezawa, S., 30, 36, 38, 64
Young People's Project, 83, 87
Yun, J. T., 30

Zeitz, P., 87
Zollman, A., 74

About the Author

Erica N. Walker is an associate professor of mathematics education at Teachers College, Columbia University. A former public high school mathematics teacher from Atlanta, Georgia, her research focuses on social and cultural factors that facilitate mathematics engagement, learning, and performance, especially for underserved students. She collaborates with teachers, schools, districts, and organizations to promote mathematics excellence and equity for young people. Her work has been published in journals such as *American Education Research Journal, Educational Leadership*, and *The Urban Review*.